Petroleum Radiation Processing

Petroleum Radiation Processing

Editor

Manoj Karkare

Petroleum Radiation Processing

Edited by **Manoj Karkare**

Printed in 2017

ISBN: 978-1-68117-035-0

Library of Congress Control Number: 2015931392

© 2016 by

SCITUS Academics LLC,
616, Corporate Way, Suite 2, 4766,
Valley Cottage, NY 10989

www.scitusacademics.com

Contents

vi

Preface

Petroleum Radiation Processing fills an information gap, providing systematic descriptions of the fundamentals of radiation-induced cracking reactions in hydrocarbons. It analyzes the basic experiments that have brought about the rapid development of radiation technology for petroleum radiation processing during the last decades. The book emphasizes high-viscous oil feedstocks that are difficult to process by conventional methods—such as heavy and high-paraffinic crude oil, fuel oil, and bitumen. It helps readers understand the mechanisms and kinetics of low-temperature radiation cracking. The book addresses the application of promising radiation methods for solving critical environmental issues, such as oil desulfurization and regeneration of used lubricants and other used oil products. Examining experimental data as well as theoretical and technical approaches, it summarizes research progress in the field of petroleum radiation processing, providing a useful reference on the theory and technology of hydrocarbon radiation processing for scientists, researchers, and students.

Editor

Chapter 1

Conversion of Isopropylbenzene over AlSBA-15 Nanostructured Materials

Francine Mota Ribeiro[1], Anne Michelle Garrido Pedrosa[2], and Marcelo José Barros de Souza[1]

[1]Department of Chemical Engineering, Federal University of Sergipe, São Cristovão, Brazil

[2]Department of Chemistry, Federal University of Sergipe, São Cristovão, Brazil

ABSTRACT

In this paper we studied the application of a series of AlSBA-15 catalysts with different Si/Al ratios on the reaction of conversion of isopropylbenzene. The catalysts were synthesized via hydrothermal method and characterized by nitrogen adsorption at 77 K and Small Angle X-ray Scattering (SAXS). Catalytic tests were accomplished in a continuous flow fixed bed microreactor. The obtained results showed catalytic activity for all AlSBA-15 catalysts producing benzene and propene as main products.

INTRODUCTION

Several technological changes have occurred in the petroleum refining industry with the aim of improving the performance of these processes, which highlights the development of catalysts that provide higher activity and selectivity in the conversion of oil derived products with high aggregate value. Mesoporous catalysts of AlSBA-15 type show a high degree of accessibility [1] , so it has been studied in cracking applications. The catalytic cracking process occurs in the presence of a catalyst which promotes the breaking of a carbon chain from long to smaller molecules, and increases the selectivity to desired products. AlSBA-15 is a mesoporous material derived from SBA-15 (Santa Barbara Amorphous). SBA-15 is a mesoporous silicate which was discovered in the 90s by Zhao and collaborators [1] . These materials have well-defined pore diameter which can be adjusted between 2 and 30 nm, and have very high BET surface area (>500 m^2/g), mesopores with hexagonally arranged, large pore wall thicknesses that result in high thermal and hydrothermal stability, greater than MCM-41 materials [2] [3] . These characteristics have involved many research fields such as adsorption, catalysis, separation, nanoscience, solid templates for other materials, among others [4]. Catalytic tests in laboratory units are typically carried out using reagent model molecules. In this paper, the isopropylbenzene was selected as model molecule, also known as cumene. This compound, an aromatic hydrocarbon (C9h12) represents oil fractions. In the catalytic, cracking of cumene typically is expected to obtain an olefin (propylene) and an aromatic (benzene) but depending on the type of catalyst and on the reaction scheme, other side reactions can occur with the obtaining of other products, which may be verified experimentally.

EXPERIMENT

The AlSBA-15 was synthesized by hydrothermal method with different atomic ratios of Si/Al (25, 50 and 75), using as structural template the Pluronic P123, tetraethyl orthosilicate as silica source, pseudobohemite as aluminum source, hydrochloric acid and distilled water as solvents. These reagents were mixed in order to obtain a hydrogel reactive with the following molar composition: 1.0 TEOS: 0.017 P123: x Al_2O_3: 5.7 HCl: 193 H_2O. Where x is 0.007 (AlSBA-15, Si/Al = 75), 0.01 (AlSBA-15, Si/Al = 50) and 0.02 (AlSBA-15, Si/Al = 25). The reagents were mixed in order to obtain a hydrogel reactive. First, the template was dissolved on the solvents under stirring and heating at 35°C. Then there was added the required amount of pseudobohemite (AlOOH) to obtain a Si/Al ratio desired. The silica source was added for last and the mixture was stirred and heated at 35°C for 24 hours to obtain a homogeneous gel. The resultant gel was transferred to an autoclave and conditioned on the oven for 48 hours at 100°C. The resulting products were vacuum filtered and washed with water, after this procedure the material has been placed to dry at room temperature for 24 hours. The SAXS of the samples were obtained at a scan angle 0° - 5° in an equipment SAXS1 of scattering X-ray small angle of the LNLS—National Laboratory Light Cyclotron in Campinas/SP/Brazil. The experiments were conducted using CuKα radiation. The survey of N_2 adsorption-desorption isotherms of the samples calcined AlSBA-15 was performed in a Quantachrome equipment NOVA1200 by nitrogen adsorption at a temperature of 77 K. The catalytic test was carried out in a fixed bed continuous flow rector manufactured by Altamira Instruments (BENCHCAT 4000 HP). The tests were conducted using a mass of 200.0 mg of catalyst. Before the start of the reaction the catalysts were activated in situ under heating in nitrogen flow of 50 mL/min starting from ambient temperature to the temperature of reaction (350°C or 400°C) under a rate heating of 5°C/min. After reaching the reaction temperature, the reactor was isolated by diverting the flow through the bypass and at that moment the syringe liquid pump was thrown which drove a syringe containing liquid isopropylbenzene by capillary flow lines of the unit. The liquid injected through the syringe (at 50 µL/min) was continuously vaporized, and being transported by the flow lines in a continuous which were heated at temperature of 200°C. This were mixing with the stream of inert gas (nitrogen) for 10 minutes until the lines could be entirely filled from the mixture, after which time being directed to the reactor in a down flow in a Weight Hourly Space Velocity (WHSV) of 3 h^{-1}. During the catalytic reactions, the reactor effluent gases are injected at time intervals of approximately 3.5 minutes, through a sampling valve Valco 6 pathways in a gas chromatograph equipped with a 6890N Agilent HP-5 column −5% Methyl Phenyl siloxane 30 m. Analysis of the products was performed using a FID detector.

RESULTS AND DISCUSSION

Catalyst Characterization

The SAXS analysis of the materials obtained showed the diffraction patterns with reflection peaks in the interval of low Bragg angle (less than 5). According to Zhao [1], three diffraction peaks are observed for SBA-15 materials which are related to the crystal planes, whose Miller indices are (100), (110) and (200). These diffraction lines are associated with a two-dimensional hexagonal symmetry P6mm typical of SBA-15 materials [2] [5]. Figure 1(a) show the SAXS patterns obtained for AlSBA-15 samples with different Si/Al ratios and in the calcined form. For the AlSBA 15 with Si/Al ratio of 25 and 50 were observed a set of diffraction peak in the region of 2θ of 0.5° to 3° which are related to(100) plane and two less intense peaks related to (110) and (200) planes, which indicating a highly ordered hexagonal material. The diffraction intensity of reflection (100) the samples mentioned above indicates that the unit cell dimensions are the same as in material SBA-15 [6].

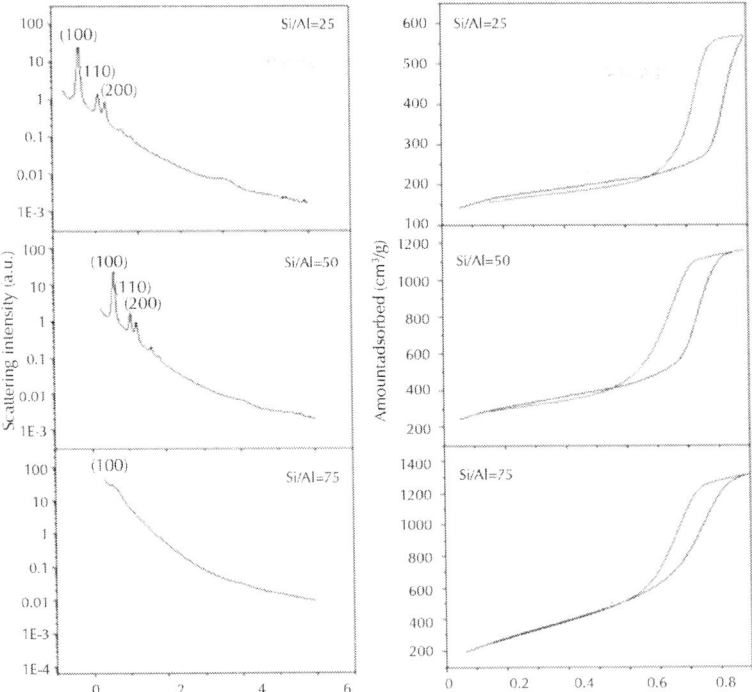

Figure 1: (a) SAXS analysis to the AlSBA-15 with different Si/Al ratios and (b) Adsorption-desorption isotherms of N2 to the AlSBA-15 with different Si/Al ratios, where: black line = adsorption and red line = desorption.

The sample with Si/Al ratio = 75 no showed no characteristic peak to this material, this means that there was no formation of the structure.

The N_2 adsorption-desorption isotherms obtained for the AlSBA-15 with Si/Al ratios of 25, 50 and 75 are shown in Figure 1(b). It can be observed that all samples exhibit adsorption/desorption isothermal of IV type in according to the IUPAC classification. The AlSBA-15 samples with Si/Al = 25 and 50 showed a "loop" of hysteresis of type I, which is associated with the porous pellets consisting of hard spherical particles of uniform size ordered regularly [7]. Hysteresis of type I are typical of mesoporous materials, thus showing that the samples AlSBA-15 with ratios Si/Al = 25 and 50 have mesoporous structure [4] [8]. The profiles of adsorption of each isotherm show a sharp increasing of the N_2 adsorbed in the range of relative pressure of 0.6 to 1.0. This suggests that the materials are very regular mesoporous channels with a narrow pore size distribution of gaussian [4] [8].

The position of the inflection points of P/Po is related to the pore size [4]. The hysteresis loop at a P/Po > 0.8 is corresponding to the inter particle capillary condensation in the pores and due to the presence of certain phases of impurities [8]. Table 1 shows the results of the textural characterization of the AlSBA-15 samples with ratios Si/Al = 25 and 50.

Catalytic Properties

The catalytic conversion of isopropyl benzene was performed at two temperatures (350°C and 400°C) in order to verify the effect of the temperature in the reactions. Figure 2(a) and Figure 2(b) shows the values of conversion as function of reaction time for different Si/Al ratios at 350°C and 400°C, respectively. At all reaction times it was observed that the conversion it was higher in temperature of 400°C than in the temperature of 350°C in the sample with Si/Al = 25. At 400°C the conversion increased slightly as a function of reaction time and after 10 minutes the conversion stabilizes. This may be related to the deposition of coke on the surface of the material.

At 350°C the conversion is maintained stabilized during the reaction time studied. Deactivation at high temperatures is common for this type of reaction where the continuous deposition of coke can inhibit the reaction. The influence of catalyst to oil ratio (C/O) in the conversion shows a similar behavior, where by analyzing the initial five minutes of reaction, it is observed that by increasing the catalyst mass, reach a maximum conversion of C/O of 2.5 and 1.3, at temperatures of 400°C to 350°C, respectively. The same is observed with temperature, and as expected for this endothermic reaction that at low pressure, in the ideal gas state, increasing the temperature favors the conversion. The parameter C/O becomes very important, because although it estimated at fixed bed conditions, is one of the commonly measured parameters during the catalytic cracking in fluidized bed reactors, where the load in a given C/O ratio is cracked in the "riser" of FCC unit, giving the idea of the proportions of catalyst and oil used for the maximization of the conversion in each fluidized bed catalytic cycle.

For AlSBA-15 catalyst with Si/Al = 50, conversion to 350°C remained relatively constant with flow time, indicating no apparent deactivation in the range examined. In this sample, the conversion at 400°C was low compared to the conversion at temperature of 350°C. The drastic reduction of the conversion seen during the initial phase of the reaction indicated rapid coke formation, which occurs on the catalyst surface. Thus, in this reaction, the decrease in conversion to the long reaction time may be due to fast blockage of acid sites by the coke formed during the reaction. The same also occurred with respect to C/O ratio, which show the conversion profile as a function of the C/O ratio to the catalyst AlSBA-15 with Si/Al = 50 with opposite behavior in comparison with AlSBA-15 with Si/Al = 25. The temperature of 350°C showed higher conversion rates in function of C/O ratio.

But this data is consistent to that found in relation to the greater catalytic activity of the AlSBA-15 sample with Si/Al = 50 at 350°C. For this sample, in both cases (temperatures of 350°C to 400°C), the best C/O ratio was about 2. For the AlSBA-15 catalysts, through gas chromatography was observed the formation of benzene and propene as reaction products.

Table 1: Textural properties of the AlSBA-15 samples

Samples	$SA_{BET} (m^2g^{-1})$	$V_p (cm^3g^{-1})$	D_{pBJH} (nm)	Wt (nm)
A1SBA-15 (25)	555	0.9	12.5	4.3
A1SBA-15 (50)	1046	1.8	8.8	7.9

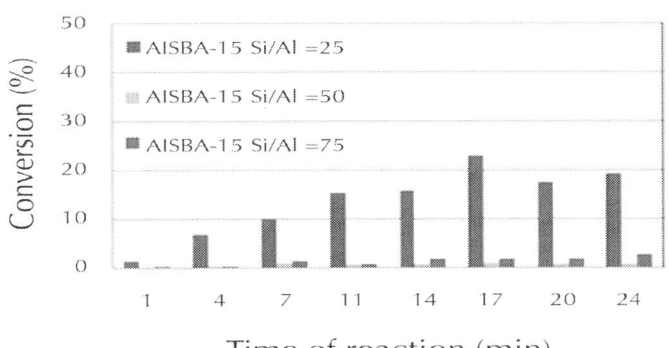

(a)

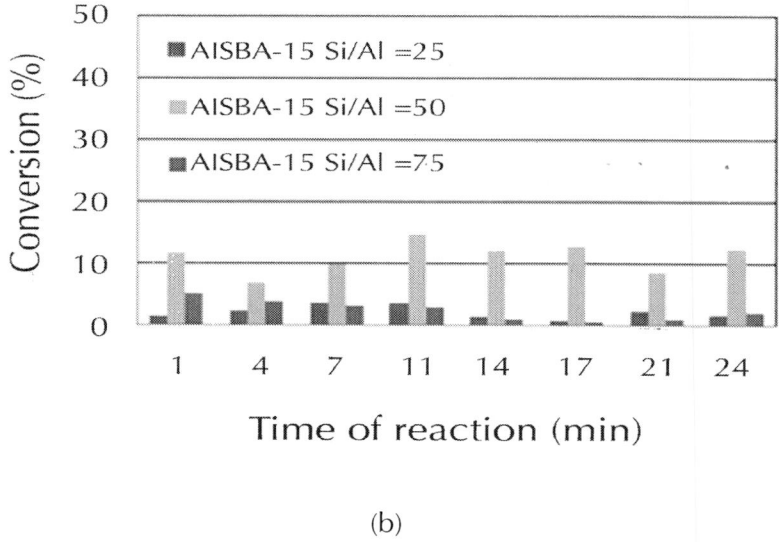

(b)

Figure 2: Conversion as function of reaction time for different Si/Al ratios: (a) at 350°C and (b) at 400°C.

To the catalyst AlSBA-15 with Si/Al = 75 the conversion of ca. 1% to 5% were observed in the studied experimental conditions. This is attributed to the high Si/Al ratio, thus making the material potentially less acidic to cracking. In this sample was not observed too the SBA-15 structure, as evidenced by SAXS analysis (Figure 1(a)).

According Trejo [9], silica-alumina prepared by sol-gel method can produce a larger quantity of Brönsted acid sites most likely due to the homogeneity of Al atoms in the silica network by which is PEO the ability to generate more Brönsted acid sites on the surface of silica-alumina by forming hydrogen bonds with the silica during the sol-gel reactions. It has been reported that the oxygen atoms in the PEO can be coordinated with cations Al_3^+ and probably coordination affects the homogeneity on silica-alumina and the local structure of aluminum networks silica which promote the generation of a greater number of Brönsted acid sites. Thus, the method of preparation of the catalyst directly influences the amount of acidic sites present in the catalyst and its catalytic behavior consecutively.

CONCLUSIONS

AlSBA-15 catalysts were satisfactorily synthesized via hydrothermal method. The characterizations via nitrogen adsorption and SAXS evidenced that samples with Si/Al of 25 and 50 presented highly ordered structure with surface area in the range of 555 to 1046 m²/g. Catalytic cracking of isopropyl benzene was studied on nanostructured materials such AlSBA-15 at different Si/Al ratios. During the reactions the formation of benzene and propene from direct cracking of isopropyl benzene model molecule was observed.

ACKNOWLEDGEMENTS

The authors acknowledge to Universidade Federal de Sergipe, with financial support from CENPES/PETROBRAS (T.C. 4600286284 da Rede Temática de Desenvolvimento de Catálise), CAPES (Coordenação de Aperfeiçoamento de Pessoal de Nível Superior) and CNPq (Conselho Nacional de Desenvolvimento Científico e Tecnológico).

REFERENCES

1. Zhao, D., Feng, J., Huo, Q., Melosh, N., Fredrickson, G.H., Chmelka, B.F. and Stucky, G.D. (1998) Triblock Copolymer Syntheses of Mesoporous Silica with Periodic 50 to 300 Angstrom Pores. Science, 279, 548-552. http://dx.doi.org/10.1126/science.279.5350.548

2. Katiyar, A., Yadav, S., Smirniotis, P. and Pinto, N.G. (2006) Synthesis of Ordered Large Pore SBA-15 Spherical Particles for Adsorption of Biomolecules. Journal of Chromatography A, 1122, 13-20. http://dx.doi.org/10.1016/j.chroma.2006.04.055

3. Gédéon, A. Lassoued, A., Bonardet, J.L. and Fraissard, J. (2001) Surface Acidity Diagnosis and Catalytic Activity of AlSBA Materials Obtained by Direct Synthesis. Microporous and Mesoporous Materials, 44, 801-806. http://dx.doi.org/10.1016/s1387-1811(01)00263-3

4. Wang, H. (1995) Magnetization and Interlayer Coupling in FeSi/

Cu Compositionally Modulated Films. Physica Status Solidi (a), 147, 529-533. http://dx.doi.org/10.1002/pssa.2211470223

5.   Kumaran, G.M., Garg, S., Soni, K., Kumar, M., Gupta, J.K., Sharma, L.D., Rama Rao, K.S. and Murali Dhar, G. (2008) Synthesis and Characterization of Acidic Properties of Al-SBA-15 Materials with Varying Si/Al Ratios. Microporous and Mesoporous Materials, 114, 103-109. http://dx.doi.org/10.1016/j.micromeso.2007.12.021

6.   Han, Y.J., Kim, J.M. and Stucky, G.D. (2008) Preparation of Noble Metal Nanowires Using Hexagonal Mesoporous Silica SBA-15. Chemistry of Materials, 12, 2068-2069. http://dx.doi.org/10.1021/cm0010553

7.   Figueiredo, J.L. and Ribeiro, F.R. (1989) Catálise Heterogênea. 1st Edition, Fundação Calouste Gulbenkian, Lisboa.

8.   Li, Y., Yang, Q., Yang, J. and Li, C. (2006) Synthesis of Mesoporous Aluminosilicates with Low Si/Al Ratios Using a Single-Source Molecular Precursor under Acidic Conditions. Journal of Porous Materials, 13, 187-193. http://dx.doi.org/10.1007/s10934-006-8003-8

9.   Trejo, F., Rana, M.S. and Ancheyta, J. (2011) Genesis of Acid-Base Support Properties with Variations of Preparation Conditions: Cumene Cracking and Its Kinetics. Industrial & Engineering Chemistry Research, 50, 2715-2725. http://dx.doi.org/10.1021/ie1008037.

Thermolysis of High-Density Polyethylene to Petroleum Products

Sachin Kumar and R. K. Singh

Department of Chemical Engineering, National Institute of Technology Rourkela, Orissa 769008, India

ABSTRACT

Thermal degradation of plastic polymers is becoming an increasingly important method for the conversion of plastic materials into valuable chemicals and oil products. In this work, virgin high-density polyethylene (HDPE) was chosen as a material for pyrolysis. A simple pyrolysis reactor system has been used to pyrolyse virgin HDPE with an objective to optimize the liquid product yield at a temperature range of 400°C to 550°C. The chemical analysis of the HDPE pyrolytic oil showed the presence of functional groups such as alkanes, alkenes, alcohols, ethers, carboxylic acids, esters, and phenyl ring substitution bands. The composition of the pyrolytic oil was analyzed using GC-MS, and it was found that the main constituents were n-Octadecane, n-Heptadecane, 1-Pentadecene, Octadecane, Pentadecane, and 1-Nonadecene. The physical properties of the obtained pyrolytic oil were close to those of mixture of petroleum products.

INTRODUCTION

Plastic materials comprise a steadily increasing proportion of the municipal and industrial waste going into landfill. Owing to the huge amount of plastic wastes and environmental pressures, recycling of plastics has become a predominant subject in today's plastics industry. Development of technologies for reducing plastic waste, which are acceptable from the environmental standpoint and are cost-effective, has proven to be a difficult challenge because of the complexities inherent in the reuse of polymers. Establishing optimal processes for the reuse/recycling of plastic materials, thus, remains a worldwide challenge in the new century. Plastic materials find applications in agriculture as well as in plastic packaging, which is a high-volume market owing to the many advantages of plastics over other traditional materials. However, such materials are also the most visible in the waste stream and have received a great deal of public criticism as solid materials have comparatively short life-cycles and usually are nondegradable.

Thermal cracking, or pyrolysis, involves the degradation of the polymeric materials by heating in the absence of oxygen. The process is usually conducted at temperatures between 500 and 800°C and results in the formation of a carbonized char and a volatile fraction that may be separated into condensable hydrocarbon oil and a noncondensable high calorific value gas. The proportion of each fraction and its precise composition depend primarily on the nature of the plastic waste and on process conditions as well.

In pyrolytic processes, a proportion of species generated directly from the initial degradation reaction are transformed into secondary products due to the occurrence of inter- and intramolecular reactions. The extent and the nature of these reactions depend both on the reaction temperature and also on the residence of the products in the reaction zone, an aspect that is primarily affected by the reactor design.

In addition, reactor design also plays a fundamental role, as it has to overcome problems related to the low thermal conductivity and high viscosity of the molten polymers. Several types of reactors have been reported in the literature, the most frequent being fluidized bed reactors, batch reactors, and screw kiln reactors [2].

Characteristics of thermal degradation of heavy hydrocarbons can be described with the following items.(1)High production of C_1s and C_2s in the gas product.(2)Olefins are less branched.(3)Some diolefins made at high temperature.(4)Gasoline selectivity is poor; that is, oil products have a wide distribution of molecular weight.(5)Gas and coke products are high.(6)Reactions are slow compared with catalytic reactions.

High-density polyethylene (HDPE) is the third-largest commodity plastic material in the world, after polyvinyl chloride and polypropylene in terms of volume. It is a thermoplastic material composed of carbon and hydrogen atoms joined together forming high-molecular-weight products. The effect of temperature and the type of reactor on the pyrolysis of HDPE has been studied, and some of the results are reviewed.

Wallis and Bhatia have done the thermal degradation of high-density polyethylene in a reactive extruder at various screw speeds with reaction temperatures of 400°C and 425°C. A continuous kinetic model was used to describe the degradation of the high-density polyethylene in the reactive extruder. It was found that purely random breakage and a scission rate which had a power-law dependence on molecular size of 0.474 best described the experimental data. The greatest discrepancy between the model prediction and the experimental data was the large molecular size region at short residence times; however, this only accounted for a very small percentage of the total distribution and was attributed to the presence of fast initiation reaction mechanism that was only significant at low conversions [3].

Conesa et al. studied the production of gases from polyethylene (HDPE) at five nominal temperatures (ranging from 500°C to 900°C) using a fluidized sand bed reactor. HDPE primary decomposition and wax cracking reactions take place inside the reactor. Yields of 13 pyrolysis products (methane, ethane, ethylene, propane, propylene, acetylene, butane, butylenes, pentane, benzene, toluene, xylenes, and styrene) were analyzed as a function of the operating conditions. From the study of HDPE pyrolysis in a fluidized sand bed reactor, they have found that the yield of total gas obtained increases in the range 500°C–800°C from 5.7 to 94.5%; at higher temperatures, the yield of total gas decreases slightly; the formation of methane, benzene, and toluene is favored by high residence times, but ethane, ethylene, propane, propylene, butane, butylenes, and pentane undergo cracking to different extents at increasing residence times and/or temperature; and the maximum yield of total gas obtained at 800°C from HDPE pyrolysis is 94.5% with the following composition: 20% methane, 3.8% ethane, 37% ethylene, 0.2% propane, 4.7% propylene, 0.3% butane, 0.4% butylenes, 2.2% pentane, 24% benzene, 2.1% toluene, 0.01% acetylene, and 0.02% xylenes and styrene [4].

Walendziewski and Steininger reported the thermal degradation of polyethylene in the temperature range 370–450°C. In the case of thermal degradation of polyethylene, an increase in degradation temperature led to an increase of gas and liquid products, but a decrease of residue (boiling point > 360°C). However, the increase of gas was not too large as compared to the sharp decrease of residue with increase of temperature. The result of analysis of gas products obtained by the pyrolysis of polyethylene at 400°C is summarized in Table 1 [1].

Table 1: Composition of gas products obtained from pyrolysis of polyethylene at 400°C [1].

Component	Thermal	Catalytic	Hydrocracking
Methane	22.7	12.4	21.1
Ethane	27.4	20.4	21.2
Ethylene	1.4	2.3	0.1

C_3	26.6	30.4	23.7
C_4	11.0	20.3	20.7
C_5	6.9	5.6	7.3
C_6	2.1	3.3	3.8

Walendziewski carried out two series of experiments of waste polymers cracking. The first series of polymer cracking experiments was carried out in glass reactor of 0.5 dm³ volume at atmospheric pressure and in a temperature range 350–420°C, the second one in autoclaves under hydrogen pressure (≈3–5 MPa) in the temperature range 380–440°C. The influence of cracking parameters, that is, reaction temperature, presence and amount of cracking catalysts, composition of the polymer feed on product yields, and composition of gas and liquid fractions is discussed. It was stated that the proper selection of the process parameters makes it possible to control, in the limited range, the product composition distribution as well as yields and composition of gas, gasoline, and diesel oil fractions [5].

Walendziewski carried out the experiments of waste polymers cracking in a continuous-flow tube reactor. The main components of the reactor unit were a screw extruder as a waste plastics feeder and a tube cracking reactor equipped with an internal screw mixer. Cracking process was realized at the temperature range 420–480°C and raw material feeding rate from 0.3 up to 2.4 kg/h. The principal process products, gaseous and liquid hydrocarbon fractions, are similar to the refinery cracking products. They are unstable due to their high olefins content (especially from polystyrene cracking), and their chemical composition and properties strongly depend on the applied feed composition, that is, shares of polyethylene, polypropylene, and polystyrene. The material balance experiments showed that the main products, liquid or solid materials in ambient temperature, contain typically 20–40% of gasoline fractions (range of boiling point 35–180°C) and 60–80% of light gas oil fractions (initial boiling point > 180°C). The solid carbon residues are similar to coal cokes and even contain 50% mineral components. Their calorific values attain 20 MJ/kg and they are solid products of quality similar to brown coals [6].

A number of studies have been reported in which a range of catalysts and reaction conditions have been employed to convert waste plastics into the hydrocarbon liquid using pyrolysis during the past four decades. The most commonly used catalysts in the catalytic degradation of high-density polyethylene are solid acids (zeolite, silica-alumina) [7–13] and spent FCC [14, 15].

This work focuses on characterization of liquid product obtained from thermal pyrolysis of virgin high-density polyethylene at different temperature ranges. Thermal pyrolysis of high-density polyethylene pellets was done in a semibatch reactor at a temperature range of 400°C to 550°C and at a heating rate of 20°C/min. The effect of pyrolytic temperature on reaction time, liquid yield, and volatiles was also studied. The obtained liquid product was characterized by different physical and chemical properties using GC-MS and FTIR.

MATERIALS AND METHODS

HDPE pellets (2.5 mm in size) obtained from Reliance Industries Ltd., India, with density 0.945 g/cc^3, Melt Flow Index (MFI) value 0.2–15 g/10 min^{-1} (at 190°C and 2.16 kg load), and melting point 133°C were used for experiments. These plastic pellets were used directly in the thermal pyrolysis reaction. The proximate analysis of HDPE pellets was done by ASTM D3173-75 and ultimate analysis was done using CHNS analyzer (Elementar Vario El Cube CHNSO). Calorific value of the raw material was found by ASTM D5868-10a.

Thermogravimetric analysis of the HDPE sample was carried out with a Shimadzu DTG-60/60H instrument. A known weight of the sample was heated in a silica crucible at a constant heating rate of 293 K/min operating in a stream of nitrogen with a flow rate of 40 mL/min from 32°C to 700°C.

The pyrolysis setup consists of a semibatch reactor made of stainless-steel tube (length: 145 mm, internal diameter: 37 mm, and outer diameter: 41 mm) sealed at one end and an outlet tube at other end as shown in the previous study [16]. The reactor is heated externally by an electric furnace, with the temperature being measured by a Cr-Al : K-type thermocouple fixed inside the reactor, and temperature is controlled by external PID controller. 20 g of HDPE sample was loaded in each pyrolysis reaction. The condensable liquid products/wax were collected through the condenser and weighed. After pyrolysis, the solid residue (wax) left out inside the reactor was weighed. Then, the weight of gaseous/volatile product was calculated from the material balance. Reactions were carried out at different temperatures ranging from 400 to 550°C.

Fourier transform infrared spectroscopy (FTIR) of the pyrolysis oil obtained at optimum condition was taken in a Perkin-Elmer Fourier transformed infrared spectrophotometer with resolution of 4 cm^{-1}, in the range of 400–4000 cm^{-1} using Nujol mull as reference to know the functional group composition. The components of liquid product were analyzed using GC-MS-QP 2010 (Shimadzu) using flame ionization detector. The GC conditions, column oven temperature progress, column used, and MS conditions are given in the Table 2.

Table 2: GC-MS conditions.

	GC-MS-OP 2010 Shimadzu	
	GC conditions	
Column oven temperature	70°C	
Injection mode	Split	
Injection temperature	200°C	
Split ratio	10	
Flow control mode	Linear velocity	
Column flow	1.51 mL/min	
Carrier gas	Helium 99.9995% purity	
	Column oven temperature progress	
Rate	**Temperature (°C)**	**Hold time (min)**
-	70	2
10	300	7.0
		(32 min total)
	Column: DB-5	
Length	30.0 m	
Diameter	0.25 mm	

Film thickness	0.25 µm	
	MS conditions	
Ion source temperature	200°C	
Interface temperature	240°C	
Start m/z	40	
End m/z	1000	

RESULT AND DISCUSSION

Proximate and Ultimate Analyses of Virgin HDPE

The proximate and ultimate analyses of virgin HDPE sample are shown in Table 3. The volatile matter is 99.9% in the proximate analysis, due to the negligible percentage of ash in virgin HDPE sample; its degradation occurs with minimal formation of residue. The oxygen is 2.51% in the ultimate analysis of virgin HDPE. The nitrogen and oxygen in the virgin HDPE sample may not be due to the fillers but due to the other ingredients which are added to resin during the manufacturing of HDPE.

Table 3: Proximate and ultimate analyses of virgin HDPE.

Properties	Virgin HDPE
Proximate analysis	
Moisture content	0.00
Volatile matter	99.92
Fixed carbon	0.00
Ash content	0.08
Ultimate analysis	
Carbon (C)	83.29
Hydrogen (H)	13.93
Nitrogen (N)	0.20

Sulphur (S)	0.07
Oxygen (O)/others	2.51
GCV (MJ/kg)	47.64

TGA and DTG Analyses of Virgin HDPE Sample

Thermogravimetric analysis (TGA) is a thermal analysis technique which measures the weight change in a material as a function of temperature and time in a controlled environment. This can be very useful to investigate the thermal stability of a material or to investigate its behavior in different atmospheres (e.g., inert or oxidizing). TGA is applied in determination of the study of thermal stability/degradation of virgin HDPE in various ranges of temperature.

From the TGA curve as shown in Figure 1, the virgin HDPE degradation started at 380°C and was completed at 510°C for a heating rate of 293 K/min in nitrogen atmosphere. The degradation temperature at which weight loss of 50% takes place was about 460°C for virgin HDPE. The temperature range for waste HDPE was 390°C to 490°C, and maximum weight loss occurred at temperature 440°C as demonstrated [16]. A similar trend of nature during the analysis of HDPE decomposition by TGA/DTG has been reported by Aboulkas et al. [19]. Differential thermogravimetry (DTG) is exactly the same as TGA, except the mass loss versus time output is differentiated automatically to give the mass loss rate versus time. Generally, both the mass loss and mass loss rate versus time are produced automatically. This is quite convenient as the rate of thermal decomposition is proportional to the volatilization or mass loss rate. Differential thermogravimetry (DTG) curve for virgin HDPE contains only one peak; this indicates that there is only one degradation step in Figure1, showing that the dominant peak is from 390°C to 510°C where the conversion takes place. Similarly, maximum weight loss occurs at 498°C from DTG profile of HDPE pellets [20].

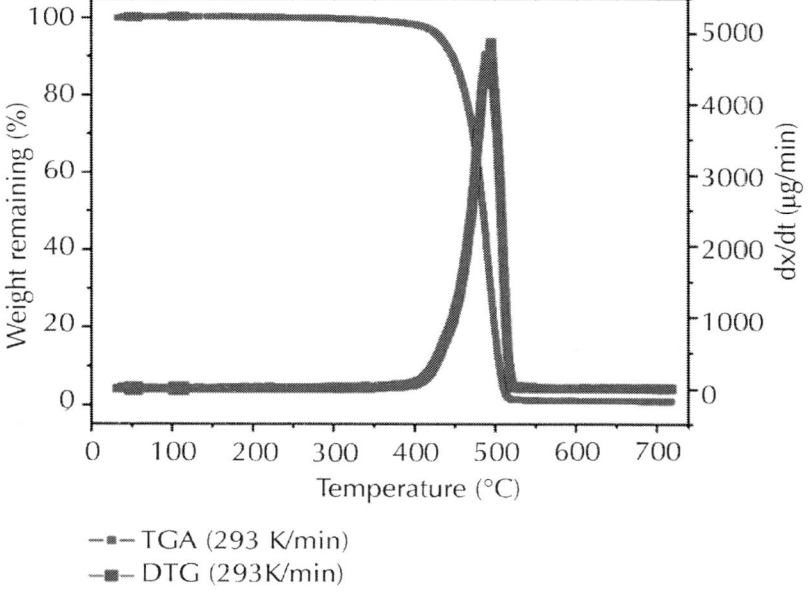

Figure 1: TGA and DTG curves of virgin HDPE.

Effect of Temperature on Product Distribution

The pyrolysis of virgin HDPE yielded four different products, that is, oil, gas, wax, and residue or coke. The distributions of these fractions are different at different temperatures and are shown in Table 4.

Table 4: Distribution of different fractions at different temperatures in thermal pyrolysis of virgin HDPE.

Temperature (°C)	Oil (wt%)	Residue (wt%)	Wax (wt%)	Gas/volatile (wt%)	Reaction time (min)
400	31.3	5.65	7.7	45.35	680
450	52.46	3.95	8.9	34.69	175
500	44.32	1.29	28.99	25.4	80
550	8.83	0.68	52.02	38.47	50

The condensable oil/wax (a mixture of alkanes that falls within the range; they are found in the solid state at room temperature and begin to enter the liquid phase past approximately 37°C) and the noncondensable gas/volatiles fractions of the reaction constituted major product as compared to the solid residue (coke) fractions. The condensable product obtained at low temperature (400°C) was low viscous liquids. With increase in temperature, the liquid became viscous/wax at and above 450°C. The hydrocarbon is continuously cracking; the wax may be representative of the intermediate-molecular-weight products. The recovery of condensable fraction was very low at low temperature, that is, at 400°C and increased with gradual increase of temperature. From Table 4, it is observed that at low temperature the reaction time was more, due to which secondary cracking of the pyrolysis product occurred inside the reactor and resulted in highly volatile product. The low-temperature molecular weight changes without volatilization are principally due to the scission of weak links, such as oxygen, incorporated into the main chain as impurities. Similarly, the low liquid yield at high temperature was due to volatilizing higher-molecular-weight products before undergoing further cracking and more noncondensable gaseous/volatile fractions due to rigorous cracking.

Effect of Temperature on Reaction Time

The effect of temperature on the reaction time of the reaction for the pyrolysis of virgin HDPE plastic is shown in Figure 2. The pyrolysis reaction rate increased and reaction time decreased with increase in temperature. High temperature supports the easy cleavage of bond and thus speeds up the reaction and lowers the reaction time. HDPE with long linear polymer chain with low branching and high degree of crystallinity led to high strength properties and thus required more time for decomposition. This shows that temperature has significant effect on reaction time and yield of liquid, wax, gaseous products, and solid residue (coke). A similar effect of temperature on reaction time for waste HDPE has been demonstrated [16].

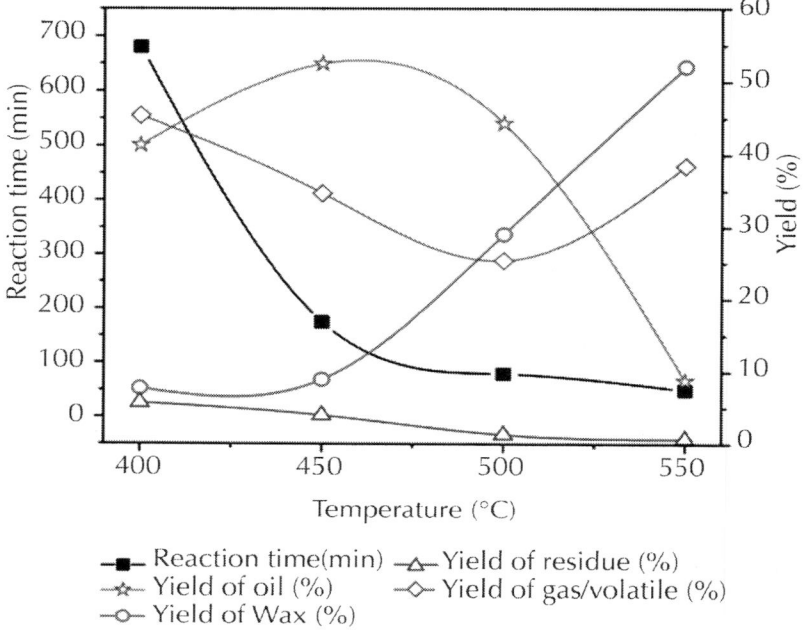

Figure 2: Effect of temperature on reaction time and product distribution.

CHARACTERIZATION OF THE LIQUID PRODUCT

FT-IR of the Oil Sample Obtained at 450°C

Fourier transform infrared spectroscopy (FTIR) is an important analysis technique which detects various characteristic functional groups present in oil. On interaction of an infrared light with oil, chemical bond will stretch, contract, and absorb infrared radiation in a specific wave length range regardless of the structure of the rest of the molecules. Figure 3 shows the FTIR spectra of virgin HDPE oil. The C–H stretching vibrations at frequency 3078.11 cm^{-1} indicate the presence of alkenes. The presence of alkanes is detected at 2918.68 cm^{-1} with C–H stretching vibrations. The C=C stretching vibrations at frequency 1647.39 cm^{-1} indicate the presence of alkenes. The presence of alkanes was detected by C–H scissoring and bending vibrations at 1440.22 cm^{-1}. The presence of alcohols, ethers, carboxylic acids, and esters is detected by C–O stretching vibrations at 907.61 cm^{-1}, and the C–H bending vibrations at frequency 719.92 indicates the presence of phenyl ring substitution bands. The results were found consistent when compared with the results of GC-MS.

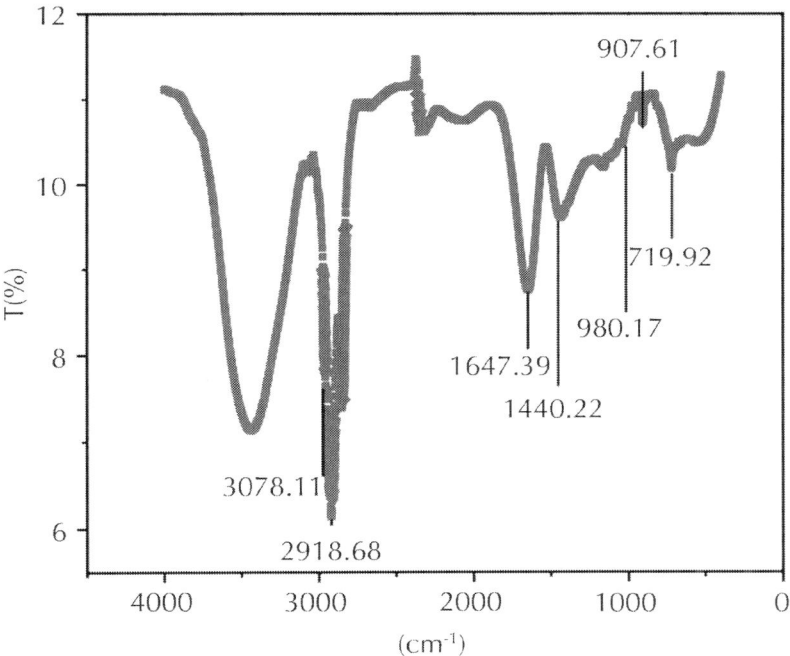

Figure 3: FT-IR spectrometry of virgin HDPE pyrolytic oil.

GC-MS of the Oil Sample

The GC-MS analysis of the oil sample obtained by the thermal pyrolysis of virgin HDPE was carried out to know the compounds present in the oil (Figure 4) and is summarized in Table 5. It has been observed that the pyrolytic oil contains around 25 compounds. Taking into account the area percentage, the highest peak areas of total ion chromatogram (TIC) of the compounds were n-Octadecane, n-Heptadecane, 1-Pentadecene, Octadecane, Pentadecane, and 1-Nonadecene. The components present in HDPE are mostly the aliphatic hydrocarbons (alkanes and alkenes) with carbon number C_{10}–C_{20}.

Table 5: GC-MS analysis of virgin HDPE pyrolytic oil.

R. time (min)	Area %	Name of compound	Molecular formula
6.301	1.24	1-Decene	$C_{10}H_{20}$
6.450	1.12	Decane	$C_{10}H_{22}$
8.105	2.04	1-Undecene	$C_{11}H_{22}$
8.238	1.78	n-Undecane	$C_{11}H_{22}$
9.735	3.50	1-Dodecanol	$C_{12}H_{26}O$
9.855	3.19	n-Dodecane	$C_{12}H_{26}$
11.541	4.62	1-Tridecene	$C_{13}H_{26}$
12.615	5.30	1-Tetradecene	$C_{14}H_{28}$
12.711	4.82	Tetradecane	$C_{14}H_{30}$
12.772	0.65	7-Tetradecene	$C_{14}H_{28}$
13.909	5.40	1-Pentadecene	$C_{15}H_{30}$
13.997	5.13	Pentadecane	$C_{15}H_{32}$
15.039	0.48	1,19-Eicosadiene	$C_{20}H_{38}$
15.130	5.36	1-Hexadecene	$C_{16}H_{32}$
15.210	5.60	n-Octadecane	$C_{18}H_{36}$
15.261	0.51	Cyclohexadecane	$C_{16}H_{32}$
16.203	0.49	1,19-Eicosadiene	$C_{20}H_{38}$
16.283	5.09	1-Nonadecene	$C_{19}H_{38}$
16.357	5.52	n-Heptadecane	$C_{17}H_{36}$
16.406	0.51	1-Heptadecene	$C_{17}H_{34}$
17.378	4.43	1-Octadecene	$C_{18}H_{36}$
17.447	5.47	Octadecane	$C_{18}H_{38}$
17.493	0.69	1-Octadecene	$C_{18}H_{36}$
18.419	3.26	1-Nonadecene	$C_{19}H_{38}$
18.482	4.67	Nonadecane	$C_{19}H_{40}$

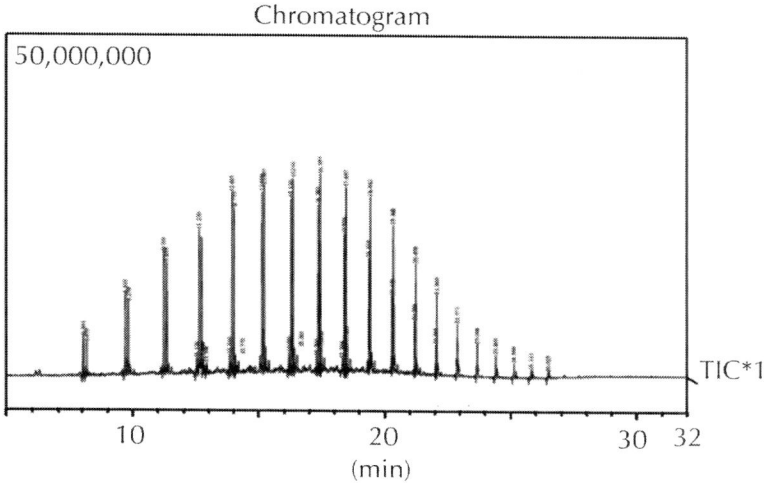

Figure 4: GC plot of oil obtained at 450°C.

Physical Properties of Oil Sample

Table 6 shows the results of physical property analysis of oil obtained from pyrolysis of virgin HDPE. The appearance of the oil is dark brownish free from visible sediments.

Table 6: Physical properties analysis of virgin HDPE pyrolytic oil.

Tests	Results obtained	Test method
Specific gravity at 15°C/15°C	0.8013	IS:1448 P:16
Density at 15°C in kg/cc	0.8006	IS:1448 P:16
Kinematic viscosity at 40°C in Cst	3.3	IS:1448 P:25
Kinematic viscosity at 100°C in Cst	1.4	IS:1448 P:25
Conradson carbon residue	<0.01%	IS:1448 P:122
Flash point by Abel method	10°C	IS:1448 P:20
Fire point	15°C	IS:1448 P:20
Cloud point	28°C	IS:1448 P:10

Pour point	18°C	IS:1448 P:10
Gross calorific value in MJ/kg	44.27	IS:1448 P:6
Sulphur content	0.03%	IS:1448 P:33
Calculated Cetane Index (CCI)	70	IS:1448 P:9
Distillation		IS:1448 P:18
Initial boiling point	72°C	
Final boiling point	364°C	

From comparison with other transportation products as shown in Table 7, the density and viscosity of liquid product can be modified by blending it with commercial transportation products. The flash point of the liquid product is in a comparable range, and a pour point of 18°C is acceptable for most geographic regions. HDPE pyrolytic oil has GCV of 44 MJ/kg which is more as compared to that of gasoline and diesel; therefore, this liquid product would perform relatively superior in engines. From the distillation report of the oil, it is observed that the boiling range of the oil is 72–364°C, which infers the presence of mixture of different oil components such as gasoline, kerosene, and diesel in the oil. The liquid product contains substantial amount of volatiles as its initial boiling point is below 100°C. From this result, it is observed that these could be possible feedstocks for further upgrading or use of lighter compounds as a diesel product.

Table 7: Product properties comparison of HDPE pyrolytic oil with commercial transportation products.

Properties	Specific gravity 15°C/15°C	Kinematic viscosity at 40°C (cst)	Flash point (°C)	Pour point (°C)	GCV (MJ/kg)	IBP (°C)	FBP (°C)	Chemical formula
HDPE pyrolytic oil	0.8013	3.3	10	18	44.27	72	364	$C_{10}-C_{20}$
Waste HDPE pyrolytic oil [16]	0.7835	1.63	1	-15	42.81	82	352	$C_{19}-C_{24}$
Gasoline [17]	0.72–0.78	-	-43	-40	42-46	27	225	C_4-C_{12}
Diesel [17]	0.82–0.85	2-5.5	53-80	-40 to -1	42-45	172	350	C_8-C_{25}
Biodiesel [17]	0.88	4-6	100-170	-3 to 19	37-40	315	350	$C_{12}-C_{22}$
Heavy product oil [18]	0.94–0.98	>200	90-180	-	-40	-	-	-

CONCLUSION

Thermal pyrolysis of virgin HDPE was performed in a semibatch reactor made up of stainless steel at temperature range from 400°C to 550°C and at a heating rate of 20°C/min. The liquid yield is highest at 450°C, highly volatile products are obtained at low temperature, and the products obtained at 500°C and 550°C are viscous liquid and wax. Reaction time decreases with increase in temperature. The functional group present in the virgin HDPE pyrolytic oil is similar to the other plastic pyrolytic oils given in several literatures. It was found that the pyrolytic oil contains around 25 types of compounds having carbon chain length in the range of $C_{10}-C_{20}$. The physical properties of pyrolytic oil obtained were in the range of other pyrolytic oils and moderate-quality products. It has been shown that a simple batch pyrolysis method can convert virgin HDPE to liquid hydrocarbon products with a significant yield which varies with temperature.

REFERENCES

1. J. Walendziewski and M. Steininger, "Thermal and catalytic conversion of waste polyolefines," Catalysis Today, vol. 65, no. 2–4, pp. 323–330, 2001.

2. S. Kumar, A. K. Panda, and R. K. Singh, "A review on tertiary recycling of high-density polyethylene to fuel," Resources, Conservation and Recycling, vol. 55, no. 11, pp. 893–910, 2011.

3. M. D. Wallis and S. K. Bhatia, "Thermal degradation of high density polyethylene in a reactive extruder," Polymer Degradation and Stability, vol. 92, no. 9, pp. 1721–1729, 2007.

4. J. A. Conesa, R. Font, A. Marcilla, and A. N. García, "Pyrolysis of polyethylene in a fluidized bed reactor," Energy & Fuels, vol. 8, no. 6, pp. 1238–1246, 1994.

5. J. Walendziewski, "Engine fuel derived from waste plastics by thermal treatment," Fuel, vol. 81, no. 4, pp. 473–481, 2002.

6. J. Walendziewski, "Continuous flow cracking of waste plastics," Fuel Processing Technology, vol. 86, no. 12-13, pp. 1265–1278, 2005.

7. Y. H. Seo, K. H. Lee, and D. H. Shin, "Investigation of catalytic degradation of high-density polyethylene by hydrocarbon group type analysis," Journal of Analytical and Applied Pyrolysis, vol. 70, no. 2, pp. 383–398, 2003.

8. J. W. Park, J. H. Kim, and G. Seo, "The effect of pore shape on the catalytic performance of zeolites in the liquid-phase degradation of HDPE," Polymer Degradation and Stability, vol. 76, no. 3, pp. 495–501, 2002.

9. G. Manos, A. Garforth, and J. Dwyer, "Catalytic degradation of high-density polyethylene over different zeolitic structures," Industrial and Engineering Chemistry Research, vol. 39, no. 5, pp. 1198–1202, 2000.

10. A. A. Garforth, Y. H. Lin, P. N. Sharratt, and J. Dwyer, "Production of hydrocarbons by catalytic degradation of high density polyethylene in a laboratory fluidised-bed reactor," Applied Catalysis A, vol. 169, no. 2, pp. 331–342, 1998.

11. S. Ali, A. A. Garforth, D. H. Harris, D. J. Rawlence, and Y. Uemichi, "Polymer waste recycling over "used" catalysts," Catalysis Today, vol. 75, no. 1–4, pp. 247–255, 2002.

12. J. F. Mastral, C. Berrueco, M. Gea, and J. Ceamanos, "Catalytic degradation of high density polyethylene over nanocrystalline HZSM-5 zeolite," Polymer Degradation and Stability, vol. 91, no. 12, pp. 3330–3338, 2006.

13. S. Karagöz, J. Yanik, S. Uçar, M. Saglam, and C. Song, "Catalytic and thermal degradation of high-density polyethylene in vacuum gas oil over non-acidic and acidic catalysts," Applied Catalysis A, vol. 242, no. 1, pp. 51–62, 2003.

14. K. H. Lee and D. H. Shin, "Catalytic degradation of waste polyolefinic polymers using spent FCC catalyst with various experimental variables," Korean Journal of Chemical Engineering, vol. 20, no. 1, pp. 89–92, 2003.

15. K. H. Lee, S. G. Jeon, K. H. Kim et al., "Thermal and catalytic degradation of waste high-density polyethylene (HDPE) using spent FCC catalyst," Korean Journal of Chemical Engineering, vol. 20, no. 4, pp. 693–697, 2003.

16. S. Kumar and R. K. Singh, "Recovery of hydrocarbon liquid from waste high density polyethylene by thermal pyrolysis," Brazilian

Journal of Chemical Engineering, vol. 28, no. 4, pp. 659–667, 2011.

17. Petroleum Product Surveys, Motor Gasoline, Summer, Winter 1986/1987, National Institute for Petroleum and Energy Research, 1986, http://www.afdc.energy.gov/afdc/pdfs/fueltable.pdf.

18. J. Tuttle and T. V. Kuegelgen, Biodiesel Handling and Use Guidelines, National Renewable Energy Laboratory, 3rd edition, 2004.

19. A. Aboulkas, K. E. Harfi, A. E. Bouadili, M. Benchanaa, A. Mokhlisse, and A. Outzourit, "Kinetics of co-pyrolysis of tarfaya (Morocco) oil shale with high-density polyethylene," Oil Shale, vol. 24, no. 1, pp. 15–33, 2007.

20. Y. Wu, A. V. Isarov, and C. O. Connell, "Thermal analysis of high density polethylene-maple woodflour composites," Thermochimica Acta, vol. 340-341, pp. 205–220, 1999.

Chapter 3

A Comparative Study on the Hydrocracking for Atmospheric Residue of Mongolian Tamsagbulag Crude Oil and other Crude Oils

Tserendorj Tugsuu[1], Sugimoto Yoshikazu[2], Byambajav Enkhsaruul[1], Dalantai Monkhoobor[1]

[1]School of Chemistry and Chemical Engineering, National University of Mongolia, Ulaanbaatar, Mongolia

[2]Energy Technology Research Institute, Advanced Industrial Science and Technology (AIST), Tsukuba, Japan

ABSTRACT

Upgrading heavy and residual oils into valuable lighter fuels has attracted much attention due to growing worldwide demand for light petroleum product. This study focused on hydrocracking process for atmospheric residue (AR) of Mongolian crude oil in the first time compared to those of other countries. Residue samples were hydrocracked with a commercial catalyst at 450°C, 460°C, 470°C for 2 hours under hydrogen pressure of 10 MPa. The AR conversion and yield of light fraction (LF) reached to 90.6 wt% and 53.9 wt%, at 470°C by the hydrocracking for atmospheric residue of Tamsagbulag crude oil (TBAR). In each sample, the yield of MF was the highest at 460°C temperature, which is valuable lighter fuel product. The polyaromatic, polar hydrocarbons and sulfur compounds were concentrated in the MF and HF because the large amount of light hydrocarbons produced from TBAR as the increasing of the hydrocracking temperature. The content of n-paraffinic hydrocarbons was decreased in HF of TBAR, on effect of hydrocracking temperature. This result suggests the longer molecules of n-paraffin (C_{20}-C_{32}) in HF were reacted better, than middle molecules of n-paraffin (C_{12}-C_{20}) in MF during the hydrocracking reaction. Because the hydrocarbon components of feed crude oils were various, the contents of n-paraffinic hydrocarbons in MF and HF of TBAR and DQAR were similar, but MEAR's was around 2 times lower and the hydrogen consumption was the highest for the MEAR after hydrocracking.

INTRODUCTION

Upgrading heavy and residual oils into valuable lighter fuels has attracted much attention due to growing worldwide demand for light petroleum product from declining reserves of sweet crude oils. Although there are large quantities of heavy oils such as atmosphericand vacuum-distilled residual oils generated as byproducts in the refinery process, it is not easy to convert these residual oils into useful hydrocarbons [1-3]. Various methods, such as thermal cracking, catalytic cracking and hydrocracking are used to produce lighter fuels from heavy oils. In these methods, large hydrocarbon molecules of residual oil have broken up into smaller and more useful hydrocarbons by cracking reaction. Cracking process is called the hydrocracking, which is reacted under the hydrogen atmosphere, with a catalyst at high temperature and pressure [4].

Even though Mongolian Zuunbayan's petroleum refinery had been closed off in 1969, geological and chemical study of Mongolian crude oils has revived in the last 2 decades. Mongolian oils are paraffinic [5-7] and have low amounts of sulfur [8], heavy metals [9], which make some troubles to the refining processes of petroleum. However Mongolian crude oils contain a large amount of atmospheric residue, which should be converted into light and middle oils in order to produce motor fuel and chemicals [10]. Although the cracking process is important for refining of heavy residue, there is no research of cracking process for the atmospheric residue of Mongolian crude oils [11]. The present research has focused on the hydrocracking for the atmospheric residue of Tamsagbulag crude oil at different temperatures, with a commercial catalyst. On comparison with those of Chinese Daqing oil and Arabian mixed Middle East oil at same conditions.

EXPERIMENTAL

Materials

Three atmospheric residues (AR) were used in this study. AR of Mongolian Tamsagbulag crude oil is coded as TBAR; Chinese Daqing—as DQAR; Arabian mixed Middle East—as MEAR. The TB oil sample was supplied by "Daqing Tamsag" Company, which is doing a mining operation in Mongolia. DQAR and MEAR were obtained from Japanese petroleum refinery. The properties of atmospheric residues are shown in Table 1.

Hydrocracking Test

Hydrocracking of AR was carried out at 450°C, 460°C and 470°C for 2 h using a fixed bed reactor that was inserted into an electric furnace with vertically shaking type. About 4 g of atmospheric residue with 200 mg of commercial catalyst was charged into the reactor, the inner volume of which was 50 ml. The reactor was pressurized by hydrogen gas up to 10 MPa at ambient temperature, and then was heated to the prescribed temperature. Reaction temperature was maintained for 2 hours in every run, and all runs were repeated two times. Table 2 shows the conditions of the hydrocracking tests.

Analysis

Initially, atmospheric residue was separated using by a distillation method to examine quantitatively its fractional composition. The hydrocracking product was handled as shown in Figure 1. At first, the toluene insoluble fraction was extracted from the hydrocracking products.

Table 1: The properties of the atmospheric residues.

Properties	Unit	TBAR	DQAR	MEAR
C	wt%	86.2	86.3	86.0
H	wt%	13.1	13.1	11.9
S	wt%	0.18	0.12	2.47
N	wt%	0.16	0.16	0.13
H/C	atom/atom	1.82	1.82	1.66
Saturate	wt%	57	59	44
Aromatic	wt%	29	33	45
Resin	wt%	9.5	5.8	7.2
C_5 Asphaltene	wt%	2.6	1.5	1.5
C_7 Asphaltene	wt%	1.4	0.3	2.6
Ni/N[a]	atom/atom	5/<1	6/1	4/10
CCR	wt%	3.3	4.6	6.4

[a]Referred to an article, which is noted on reference [9]

Table 2. The condition of hydrocracking test.

Sample	Name	TBAR, DQAR, MEAR
	Weight	4.0 g
Catalyst	Type	NiMo/Al$_2$O$_3$
	Weight	0.2 g
H$_2$ Gas	Pressure	10 MPa
	Temperature	450°C, 460°C, 470°C
	Retention time	2 hours

The toluene soluble fraction was recovered after solvent evaporation, then divided into four fractions by a distillation method: light fraction (LF) boiling point less than 220°C, middle fraction (MF) boiling point of 220°C - 350°C, heavy fraction (HF) boiling point of 350°C - 500°C and bottom boiling point over than 500°C. The separated distinct fractions were weighed to check a material balance including the yield of gases products.

After the reaction, gases products were subjected to Gas chromatography coupled with thermal conductivity detector (GC-TCD; Agilent, 6890) to estimate its composition. Then, contents of methane, ethane and propane gases were calculated using a calibration of standard gas samples. The Gas chromatography system coupled with a sulfur chemiluminescence detector (GC-SCD; Agilent 6890) was used to determine the sulfur content in liquid products from hydrocracking. The distillation curve of the liquid product was examined using a GC-FID system (Agilent, model 6890GC) that was equipped with a fused silica column 5 m long.

Equations

We used the following equations to calculate the conversion of AR (1), hydrogen consumption (2) and the yield of light fraction (3).

$$C_{AR} = \frac{W^0 - W^1}{W^0} \times 100\%$$

(1)

C_{AR}—Convertion of AR, wt%;

W^0—Initial weight of atmospheric residue, g;

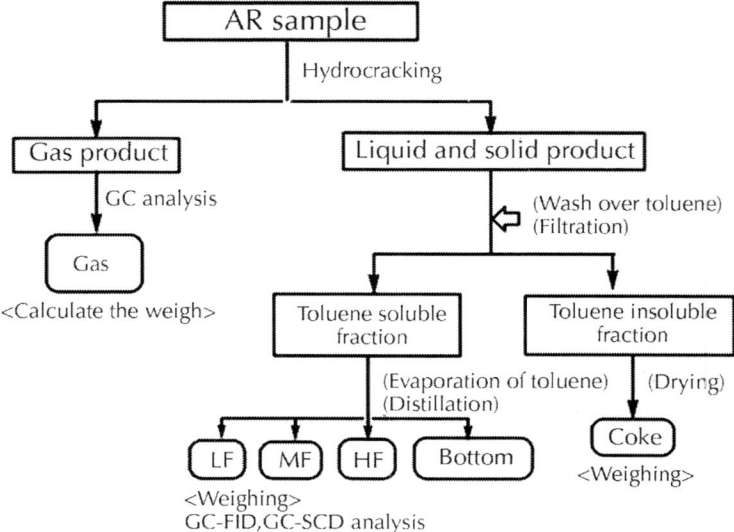

Figure 1: Product separation method of AR hydrocracking test (AR-Atmospheric residue of crude oil, LF-Light fraction, <220°C, MF-Middle fraction, 220°C - 350°C, HF-Heavy fraction, 350°C - 500°C, Bottom-Residue, >500°C).

W^1—Weight of atmospheric residue after the reaction, g;

$$W_{H2} = \frac{W_{H2}^0 - W_{H2}^1}{W^0} \times 100\%$$

(2)

W_{H2}—Hydrogen consumption, wt%;

W_{H2}^0—Weight of hydrogen, which was charged into reactor, g;

W_{H2}^1—Weight of hydrogen after the reaction, g;

W^0—Initial weight of atmospheric residue, g.

$$W_{LF} = \left(W_{AR} + W_{H2} \right)$$
$$- \left(W_{Gas} + W_{MF} + W_{HF} + W_{Bot} + W_{Coke} \right) \tag{3}$$

According to the feature of the product separation method we calculated the yield of LF (3) from the material balance, which was limited from 100.4 wt% to 101.9 wt%. The excess of 100 wt% was provided by the amount of hydrogen consumption [12].

RESULTS AND DISCUSSIONS

Table 3 shows the product distribution after the hydrocracking of AR samples at different temperatures for 2 hours under hydrogen pressure of 10 MPa, using a fixed bed reactor.

The yield of LF reached to 53.9 wt%, the yield of liquid fractions (<350°C) including gas product reached to 88.9 wt% in the hydrocracking for TBAR at the highest temperature of 470°C. Also the AR conversion was increased from 56.7 wt% to 90.6 wt%, when the hydrocracking temperature increased from 450°C to 470°C. The amount of hydrogen consumption was the highest for MEAR. It should be explained by the lowest H/C atomic ratio of corresponding feedstock. The H/C ratio of MEAR was the lowest (1.66) as shown in Table 1.

Figure 2 summarized a dependence of each product yield on reaction temperature for AR samples. It was evident that the hydrocracking product was lightened as the increasing of reaction temperature. Hydrocracking reaction of all AR samples used in this research at the temperature of 470°C provided the largest amount of gas and LF, consequently the lowest yield of HF and bottom. The yield of MF was the highest at temperature of 460°C in all of AR samples, however the variation for yield of MF was not so high compared to the yields of the other products, by the increasing of reaction temperature.

The yields of gas products after the catalytic hydrocracking of the AR samples are illustrated in Figure 3. With the TBAR the highest yield of gases product was produced after the run of hydrocracking at the temperatures of 450°C and 470°C. Also the methane content in gases products was the lowest, but the propane content was the highest after every run of the hydrocracking for AR samples. The ratio of the contents of C_1-C_3 gases was nearly constant for the all of AR [13].

Table 3: The product distribution after hydrocracking of atmospheric residue samples at different temperatures.

Samples	Initial fractional composition (wt%)		Temperature (°C)	Hydrogen consumption (wt%)	AR conversion (wt%)	Content of products (wt%)					
	HF	Vacuum residue				Gas	LF	MF	HF	Bottom	Coke
TBAR	55.8	44.2	450	0.5	56.7	3.7	29.3	22.0	38.0	5.3	2.2
			460	0.4	74.9	5.7	43.0	26.1	23.6	1.5	0.5
			470	0.9	90.6	11.4	53.9	23.6	8.6	0.8	2.6
DQAR	31.7	68.3	450	0.4	47.9	2.8	27.9	14.8	30.8	21.3	2.8
			460	0.4	69.9	5.0	37.9	24.8	23.4	6.7	2.6
			470	0.9	85.8	10.3	53.2	21.4	12.5	1.7	1.8
MEAR	33.6	66.4	450	1.2	46.2	2.8	24.3	18.3	38.5	15.3	2.0
			460	1.1	79.0	6.1	47.9	24.2	19.3	1.7	1.9
			470	1.9	90.2	9.3	60.0	19.9	8.5	1.3	2.9

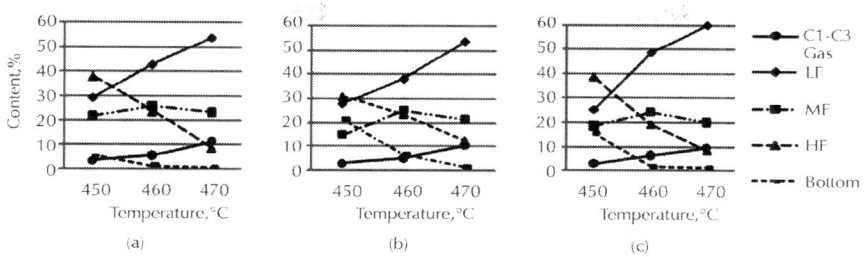

Figure 2: Relationship between the content of products and the temperature of hydrocracking. (a) TBAR; (b) DQAR; (c) MEAR.

The contents of subfractions, hydrocarbons and the amount of sulfur in MF and HF after the hydrocracking of AR samples were shown in Table 4. The content of (<254°C) subfraction was decreased, in place of it, the contents of (<344°C), (>344°C) subfractions in MF were expanded as the increasing of reaction temperature of the hydrocracking. Also the content of (<344°C) subfraction was decreased, in place of it, the contents of (<496°C) subfraction in HF was expanded as the increasing of reaction temperature. It means that the hydrocarbons component became heavier in the MF and HF after the hydrocracking of all of AR samples as the increasing of reaction temperature.

The content of saturate hydrocarbons was decreased, but the contents of polyaromatic, polar hydrocarbons and the amount of sulfur compound were expanded in MF of TBAR as the increasing of temperature for hydrocracking. Also the amount of sulfur in HF was increased, by dependent of the reaction temperature. Summation of this result and the product distribution after hydrocracking of AR (Table 3), polyaromatic, polar hydrocarbons and sulfur compounds were concentrated in MF and HF because the large amount of light hydrocarbons produced from AR and moved to LF as an increasing of the reaction temperature [14].

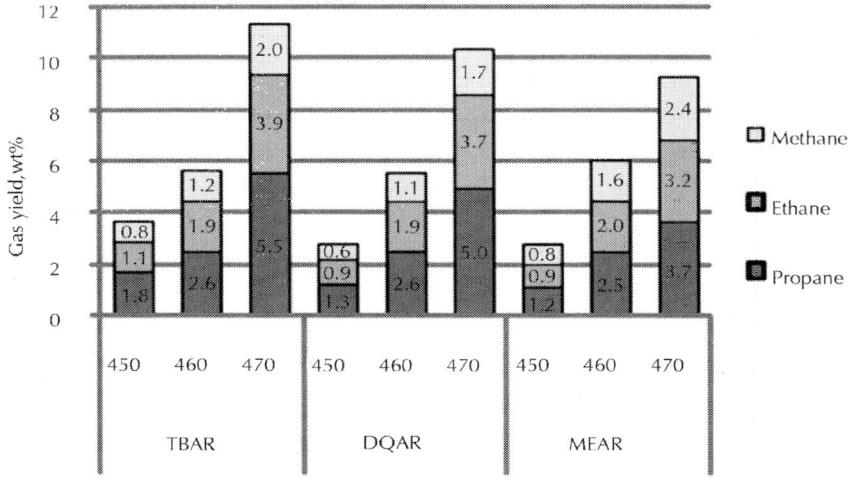

Figure 3: The yield of C_1-C_3 gases after the hydrocracking.

Table 4: Contents of subfractions, hydrocarbons and the amount of sulfur in hydrocracking products of AR samples (wt%).

Sample Temperature Properties	TBAR			DQAR			MBAR		
	450°C	460°C	470°C	450°C	460°C	470°C	450°C	460°C	470°C
Subfractions	In middle fraction (MF)								
<151°C	-			-	0.1	0.1	0.1	0.2	0.2
<254°C	42.8	32.6	29.7	46.8	40.9	35.4	36.0	33.4	27.9
<344°C	53.9	60.1	61.9	48.1	54.9	59.6	59.5	60.1	63.6
>344°C	3.3	7.3	8.4	5.1	4.1	4.9	4.4	6.3	8.3
Content of hydrocarbons									
Saturates	78.7	76.3	69.1	79.1	79.3	74.6	66.2	62.7	51.5
Mono-aromatics	18.3	17.5	19.7	16.9	17.0	18.0	28.9	29.3	32.4
Polyaromatics	2.8	5.5	9.9	2.0	3.0	6.3	4.8	7.8	15.7
Polar	0.2	0.7	1.3	0.7	1.1	2.0	0.1	0.2	0.3
n-paraffins	46.3	46.1	45.9	46.7	46.5	45.0	21.2	26.1	25.9
Sulfur	19	29	35	9	14	21	180	330	750
Subfractions	In heavy fraction (HF)								

<254°C	0.2	1.2	0.5	0.2	1.7	1.5	1.4	0.9	4.7
<344°C	343	213	13.4	27.3	30.3	14.4	22.3	23.4	15.1
<496°C	63.4	71.3	81.0	70.8	65.4	78.3	72.4	72.6	76.6
>496°C	2.1	6.2	5.1	1.7	2.6	5.8	3.9	3.1	3.6
n-paraffins	41.1	373	29.4	40.4	37.7	32.4	18.6	20.5	14.8
Sulfur	22	125	200	36	120	140	980	1160	2320

The content of n-paraffinic hydrocarbons in HF was decreased by dependent of the reaction temperature. However the content of the n-paraffinic hydrocarbons was not changed in MF of TBAR. This result suggests the longer molecules of n-paraffin (C_{20}-C_{32}) in HF were reacted better, than middle molecules of n-paraffin (C_{12}-C_{20}) in MF during the hydrocracking reaction [15]. The contents of n-paraffins in MF and HF of TBAR and DQAR were similar, but MEAR's was around 2 times lower after hydrocracking because the hydrocarbon component of those AR samples and feed crude oils were the various [12,13].

CONCLUSIONS

Atmospheric residue of Mongolian Tamsagbulag crude oil (TBAR) was tested for hydrocracking in the first time. In order to compare a hydrocraking reactivity of TBAR sample with Chinese Daqing (DQAR) and Arabian mixed Middle East (MEAR) samples were tested with commercial NiMo/Al$_2$O$_3$ catalyst at different temperatures of 450°C - 470°C. Some conclusions can be drawn as follows:

1) In the hydrocracking of TBAR, the yield of liquid fractions including of gas product (<350°C) expanded sharply as the increasing of reaction temperature. The yield of middle fraction (MF) from TBAR was the highest at temperature of 460°C;

2) The polyaromatic, polar hydrocarbons and sulfur compounds were concentrated in MF and HF of TBAR because the large amount of light hydrocarbons produced from AR and moved to LF as the increasing of the reaction temperature;

3) With the hydrocracking of TBAR and DQAR, the content of the n-paraffinic hydrocarbons more decreased in HF than MF by dependent of the reaction temperature. This result suggests the longer molecules of n-paraffins (C_{20}-C_{32}) in HF were reacted better, than the middle molecules of n-paraffin (C_{12}-C_{20}) in MF during the hydrocracking reaction of TBAR, DQAR samples;

4) The contents of n-paraffinic hydrocarbons in MF and HF of TBAR and DQAR were similar, but MEAR's was around 2 times lower than those and the hydrogen consumption was the highest after hydrocracking for the MEAR, because the hydrocarbon component of those AR samples and feed crude oils were various.

ACKNOWLEDGEMENTS

The authors gratefully acknowledge the opportunity to carry out this work by the Advanced Fuel Group, Energy Technology Research Institute (ETRI), National Institute of Advanced Industrial Science and Technology (AIST), Japan.

REFERENCES

1.   B. Enkhsaruul and Y. Ohtsuka, "Cracking Behavior of Asphaltene in the Presence of Iron Catalysts Supported on Mesoporous Molecular Sieve with Different Pore Diameters," Fuel, Vol. 82, No. 13, 2003, pp. 1571-1577. doi:10.1016/S0016-2361(03)00094-2

2.   E. Fumoto, A. Matsumura, S. Sato and T. Takanohashi, "Kinetic Model for Catalytic Cracking of Heavy Oil with a Zirconia-Alumina-Iron Oxide Catalyst in Stream Atmosphere," Energy & Fuels, Vol. 23, No. 11, 2009, pp. 5308-5311.doi:10.1021/ef9006164

3.   M. Kouzu, Y. Kuriki, K. Uchida, K. Sakanishi, Y. Sugimoto and I. Saito, "Catalytic Hydrocracking of Petroleum Residue over Carbon-Supported Nickel-Molybdenum Sulfide," Energy & Fuels, Vol. 19, No. 3, 2005, pp. 725-730. doi:10.1021/ef049895h

4.   V. Simanzhenkov and R. Idem, "Crude Oil Chemistry," Marcel Dekker, New York, 2003.

5.   B. Shirchin, E. Nordov, D. Monkhoobor and A. Sainbayar, "Study on Main Physical and Chemical Characteristics of East Mongolian Petroleum," Journal of Industrial and Engineering Chemistry (Korea), Vol. 14, No. 4. 2003, pp. 423-425.

6.   E. Enkhtsetseg, B. Byambagar, D. Monkhoobor, B. Avid and A. Tuvshinjargal, "Determination of Sterane and Triterpane in the Tamsagbulag Oilfield," Advances in Chemical Engineering and Science, Vol. 1, No. 3, 2011, pp. 163-168.doi:10.4236/aces.2011.13024

7.   B. Khongorzul, "Feature of the Hydrocarbon Composition and High Molecular Compounds of the High Paraffinic Oil from Mongolia," Ph.D. Thesis, Russian Academy of Science, Tomsk, 2008, pp. 67-77.

8. B. Shirchin, E. Nordov, D. Ganzorig and Ts. Tugsuu, "Study Review in the Fields of Mongolian Oil Chemical Technology for Last 10 Years," The Sustainable Development of Mongolia and Chemistry International Symposium, Ulaanbaatar, 11-14 September 2002, pp. 25-27.

9. J. Sainbayar, D. Monkhoobor and B. Avid, "Determination of Trace Elements in the Tamsagbulag and TsagaanEls Crude Oils and Their distillation Fractions Using by ICP-OES," Advances in Chemical Engineering and Science, Vol. 2, 2012, pp. 113-117. doi:10.4236/aces.2012.21013

10. A. Sainbayar, E. Nordov and D. Monkhoobor, "Comparison of Hydrocarbon's Composition of Main Oil Fractions between Tamsagbulag and Zuunbayan Oils, Mongolia," In: K. L. Montclaire, Ed., Petroleum Science Research Progress, Nova Science Publisher Inc., New York, 2008, pp. 329-349.

11. Ts. Tugsuu, Y. Sugimoto and B. Enkhsaruul, "Catalytic Hydrocracking for Atmospheric Residue of Mongolian and Other Crude Oils," Proceedings of 4th International Conference on Chemistry, Green Chemistry and Advanced Technology, 7-9 October 2010, pp. 215-219.

12. Y. Sugimoto, Y. Aihara, A. Matsumura, A. Ohi, S. Sato and I. Saito, "Processing of Middle East Crude with Canadian Oil Sands Bitumen-Derived Synthetic Crude Oil," Journal of the Japan Petroleum Institute, Vol. 49, No. 1, 2006, pp. 1-12.

13. Y. Sugimoto, "Slurry Phase Hydrocracking of Heavy Oil over Ni-Mo/Carbon Catalyst," 16th Saudi Arabia-Japan Joint Symposium, Al Khobar, 5-6 November 2006.

14. T. Kabe, A. Ishihara and W. Qian, "Hydrodesulfurization and Hydrodenitrogenation: Chemistry and Engineering," Wiley-Vch, Weinheim, 1999.

15. B. Enkhsaruul and Y. Ohtsuka, "Hydrocracking of Asphaltene with Metal Catalysts Supported on SBA-15," Applied Catalysis A: General, Vol. 252, No. 1, 2003, pp. 193-204.doi:10.1016/S0926-860X(03)00469-1

Chapter 4

Bioenergy Sources and Representative Case Studies in Mexico

Gibrán S Alemán-Nava[1], Luisaldo Sandate-Flores[1],
Alexander Meneses-Jácome[2], Rocío Díaz-Chavez[3], Jean-
Francois Dallemand[4] and Roberto Parra[1]*

[1]Department of Environmental Bioprocesses, Water Center for Latin America and the Caribbean, Mexico
[2]Environmental Engineering, St. Thomas University, Bucaramanga, Colombia
[3]Centre for Environmental Policy, Imperial College London, London SW7 1NA, UK
[4]European Commission, Joint Research Centre, Institute for Energy, Via E. Fermi 2749, TP 450, 21027 Ispra (Va), Italy

ABSTRACT

Energy policies during the last years have tried to promote the use and development of renewable energy, with bioenergy representing the highest potential. It is estimated that this source could supply 40% of primary energy consumption in Mexico, with a potential production of 3,569 PJ/year. Attempts to exploit biomass energy in Mexico have encouraged the development of various technologies, focused mainly in the production of biogas, biodiesel and improvement of wood stoves and charcoal furnaces. These technologies have relevant projects throughout the country. Biogas has been generated using landfills, swine waste and wastewater. Biodiesel has been produced from different sources such as, oleaginous crops, recycled oil or animal tallow. Finally, since the use of firewood as heating and cooking source is widely used in Mexico (27 million people), development of this technology has also played an important role in the exploitation of bioenergy as well as the improvement of charcoal furnaces. This paper presents an overview of the potential of bioenergy potential and presents main case studies in Mexico.

INTRODUCTION

The Energy Reform, recently approved by the Congress of the Union at the end of 2013, has allowed the creation of a fund, after covering 4.7% of GDP based on the year 2013, to strengthen renewable energy sources. Specifically, 10% of this fund will be used to finance projects in science, technology and renewable energy sources, and 10% will fund scholarships for development of human capital through postgraduate formation; however roadmaps to define energy production from each renewable energy are still missing [1]. Four main legal instruments are expected to promote renewable energy in Mexico. One is the recent Energy Reform approved by the Congress of the Union. The second instrument is the General Law for Climate Change adopted in May 2012, which sets the goal of 35% of energy generated in the country should come from renewable sources by 2024 [2]. The third legal instrument is the law of Promotion and Development of

Bioenergy in order to achieve energy diversification and sustainable development in the production of energy [3,4]. Finally, the law for the Use of Renewable Energy and Finance of the Energy Transition recently modified and approved [5]. This law establishes, among other issues, the legal aspects and conditions for the use of renewable energy and clean technologies as well as reducing the use and dependency of fossil fuels. For instance, adding 2% of ethanol to gasoline in Guadalajara, Monterrey and Mexico City was introduced at the end of 2012 [6]. These four legal instruments are expected to create a better framework to support renewable energy in general and also a future green economy in Mexico.

According to the Global Status Report 2013 [7], the global energy consumption in 2011 was estimated at 471.8 exajoules (EJ), with fossil fuels supplying 78.2%, nuclear power supplying 2.8%, and renewable energy supplying about 19% of the global final energy demand; this is divided as follows 9.7% came from modern renewable sources, including hydropower, wind, solar, geothermal and biofuels. Traditional biomass, which is used primarily for cooking and heating in rural areas of developing countries, and could be considered renewable, accounted for approximately 9.3% of the total final energy demand. According to the national energy databases [8], Mexico produced 219.5 million tons of oil-equivalent energy during 2011. An estimated 88.7% came from fossil fuels, 6.9% from renewable sources, 3.17% from charcoal and the remaining 1.3% from nuclear sources (Figure 1). The use of biomass as a primary source of bioenergy has been decreasing in Mexico since 1965, when it constituted 15.3% of the total primary energy supply. This share represented only 5.3% in 2005 [8]. This is mainly because firewood is mainly used by rural communities in the country [9] who use it for cooking and heating purposes. However, the rural population has decreased from almost 50% in 1960 to 21% in 2013 [10,11].

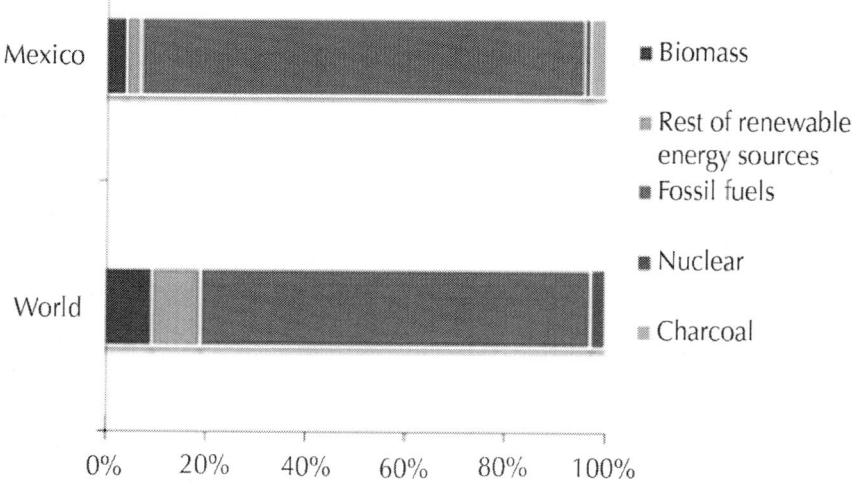

Figure 1: Energy share of global and Mexico final energy consumption in 2011.

Almost three quarters of the publications on renewable energy in Mexico, during the last 30 years, have focused on biomass research [1]. These publications have tackled different topics including management of woody biomass [12,13], biofuels (biodiesel, bioethanol) [14-16], biogas production [17,18], greenhouse gas mitigation from biomass utilization [19], biomass stoves [20-22], agricultural residues [23] among others. However, it is hard to find a paper which includes successful studies on a large scale in the country. In this context, this work gathers important information from studies, published articles and national reports about bioenergy potential and presents main case studies in Mexico.

BIOENERGY POTENTIAL IN MEXICO

Worldwide, Mexico was ranked sixth, as crude oil producer, and eighteenth as natural gas producer in 2009 [24]. Although Mexico is a producer of crude oil, there are serious problems regarding the supply of finished petroleum products. Primarily, the origin of these problems comes from the lack of investment to increase refining capacity in the country and, also from the lack of diversification of energy sources [4]. Previous studies show that bioenergy may be increased substantially in order to reach up to 16.17% of Mexico's total energy supply for electricity generation, transportation and rural residential sectors by 2030 [25,26]. It is also important to point out that Mexico is the third largest country in Latin American and the Caribbean in terms of cropland area [27]. By 2006, it was estimated that Mexico had 75.3 million tons of dry matter from crop residues that potentially could be transformed into bioenergy, Valdez-Vazquez et al. detected thirty five Mexican municipalities with a high estimated bioenergy potential (280,320 to 2,181,021 tons of dry matter per year) [26]. The Mexican Ministry of Energy estimated that the potential energy per year in 2030 would be 85,500-119,498 MW from biomass, but in 2013 the installed capacity was only 645 MW in the country [28].

The scenarios for mitigation of emissions show that bioenergy can play a key role in the energy supply in the medium and long term. The German Advisory Council on Global Change (WBGU, for its initials in German) estimated the potential of bioenergy in the global energy system to be between 80 and 170 EJ, equivalent to 17%-36% of primary energy consumption in the world in 2008 [29]. There is great potential in biomass as energy resource. In a detailed study on the reduction of carbon emissions in Mexico, financed by the World Bank, the energy potential of the main sources of bioenergy available in the country [30] was evaluated (Figure 2) as equivalent to 3,569 PJ/year, or 42% of the country's primary energy consumption in 2008. It is important to point out that these estimations were based on suitable lands for each crop mentioned in Figure 2 and excluded those a) used for agriculture, b) covered by forests, jungles and other natural hedges, c) belong to conservation areas and d) non arable because they have a slope higher than 4-12% [31].

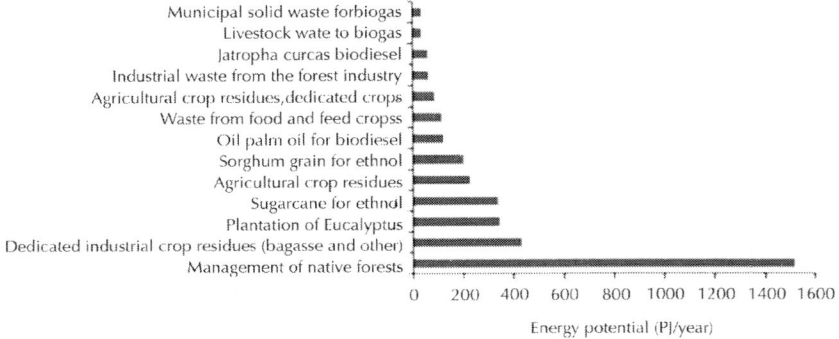

Figure 2: Biomass energy potential divided by sources in Mexico.

CASE STUDIES IN MÉXICO

There are several pathways for direct combustion, gasification, fermentation and anaerobic digestion allowing the use of biomass as a sustainable energy source [32]. There is significant experience in the area of bio-digesters, methane capture and electricity generation in landfills [17], as well as efficient wood stoves for cooking in rural areas in Mexico. Furthermore, there are emerging initiatives for liquid biofuels, particularly biodiesel, and also research groups working in materials and processes for biofuels for first, second and third generation [33,34]. Different successful bioenergy cases were researched in Mexico for big and small scales, as mentioned below.

Biogas

Gaseous biofuels are classified according to their procurement processes. Such processes may be biological, to generate biogas or thermochemical to produce syngas (synthesis gas). Biogas can be used as fuel in stoves and boilers; for domestic lighting and heating, and to fuel internal combustion engines or gas turbines, generating power or electricity. In the case of Mexico, the country has relevant experience on biological methods for biogas production (Table 1).

Table 1: Projects for biogas production.

Project	Location	Start date	Capacity	Source of energy
Biogas from landfill	Salinas Victoria, Nuevo León	2009	7.4 MW (2003), 12.72 MW (2007), 15.9 MW (2010)	Biogas from muynicipal waste
Biogas from swine waste	Cadereyta, Nuevo Leon	2005	65 kW	Pig farm manure
Biogas from wastewater treatment	León, Guanajuato	2008	2 plants of 55 kW	Cattle manure

Biogas from Landfill

Bioenergy Nuevo Leon, SA de CV (BENLESA) is the first project in Mexico and Latin America using biogas as renewable energy produced from a landfill. BENLESA is the result of a partnership between private enterprise Bioeléctrica Monterrey, SA de CV and the government of the State of Nuevo León, through the System Management and Processing of Organic Waste (SIMEPRODE), a public agency, and Bioeléctrica Monterrey and International Energy Systems, S.A. de C.V. (SEISA). These companies are completely Mexican, subsidiaries of the Group Gentor. SEISA has extensive experience in the area of generating clean energy through cogeneration projects [35].

Biogas from Swine Waste

The Agricultural Development Corporation of Nuevo León opened in August 2006 on the farm "El Chancho", Cadereyta, the first of three electricity generating plants using biogas. The initiative includes a set of nine pig farms in the municipalities of Cadereyta, Allende and Montemorelos. These farms have mitigated their methane emissions to the atmosphere through the capture and burning of biogas generated by digestion of swine manure. The project began as a response to the problem of manure disposal, the impact on production costs because ofthe large increase in volume due to water during the rainy season. It seeks to use pig manure for: a) obtaining biogas, b) generate electricity for consumption, c) sell surplus energy to the national grid (CFE), and d) obtain usable organic sludge as an agricultural fertilizer [29,36].

Biogas from Wastewater Treatment

The proposal raised to address the need to reduce Biological Oxygen Demand (BOD) and thus reduce the energy cost of the aerobic phase [38] of wastewater treatment plants from slaughterhouse "TIF 333". For the slaughterhouse, the amount of biogas produced was not a priority, but there was the possibility of using it as energy in the boiler, consuming up to 150 m³/d. Pretreatment reduces the incoming BOD from 4,700 ppm O_2 to 2012 ppm O_2 at the outlet. In the final issue of 612 ppm BOD, O_2 is achieved. Although 150 m³ of biogas/d is used in the boiler, the total generation is 157 m³ of biogas/d [29,37].

Efficient Wood Stoves

About 2 billion people worldwide rely on biomass for cooking and heating. In Mexico, the use of firewood represents about 10% of primary energy and contributes 46% of the energy demanded by the residential sector [39]. In Mexico, firewood is used by over 27 million people, 89% of the rural population uses wood as the main fuel for cooking, while in urban areas firewood users represent 11% of the population [40]. There are various projects developed in Mexico (Table 2).

Table 2: Projects for use of wood stoves.

Project	Location	Start date	Size	Characteristics
Patsari stove	Michoacán and 15 more states	2003	105 cm X 70 cmX27 cm	The exterior is made of brick which includes an optimized combustion chamber and tunnels to reduce the production of emissions.
Mexalitl stove	Nuevo Leon, Chihuahua, State of Mexico and Yucatan	2008	35 cmX70 cm	The exterior is made of concert with a combustion chamber of ceramic.
ONIL stove	Chiapas, Oaxaca, Veracruz, Puebla, Hidalgo, State of mexico, Queretaro and Guanajuato	2010	80 cmX54 cmX20 cm	Stove made from cement, sand and iron with a volume of 15 L.

Patsari Stove

The Interdisciplinary Group for Appropriate Rural Technology A.C. in collaboration with the Center for Ecosystem Research has promoted, since 2003, the use of the very efficient wood stove "Patsari" as a means to reduce adverse impacts to health and the environment, and promoting sustainable use of firewood in Mexico, and thereby improving the quality of life of rural families. Patsari means "the one who keeps" in the Purépecha language, referring to the way it works, storing heat, and so preserving user's health and sustainability of forests. It is constructed in situ, combining the use of both local and commercial materials. It achieves high efficiency, resulting in higher levels of adoption among users, shorter construction times and durability. The project is implemented in Michoacán and 15 other Mexican states, and it has 5 main components: a) innovation and technology development; b) dissemination of efficient stoves; c) development of local small businesses; d) monitoring and impact assessment e) strengthening the program. It has been quantified that each Patsari stove can mitigate between 3.0 and 8.5 tCO_2 eq/year, depending on the user type and the renewability of the firewood used [20,41].

Mexalit Stove

This stove has a weight of 76 kg, thus facilitating its distribution. The volume of the chamber is 15 L for a capacity of up to 3.5 kg of wood. It was designed to reduce from 50% to 60% fuel consumption, compared to traditional open stoves, and to improve combustion heat. The main burner is located at the center point of the generation of the flame. The chamber height generates maximum use of the flame at any time, allowing the heat to continue once the flame is extinguished due to the accumulated heat inside the stove, since the interior is lined with mud-manufactured bricks [42].

Onil Stove

This stove was designed using the principles of combustion "Rocket elbow" developed by Dr. Larry Winiarski of Aprovecho Research Center, Oregon. These principles have been used to create many ecological stoves currently used in the world. The stove and each of its components were tested for one year in the community of Santa Avelina (Guatemala) to ensure that needs have families were covered and at the same time met the requirements of quality, functionality and durability. Permanent adjustments and improvements have been made to ensure technology acceptance and adoption. There have been so far more than 16,720 stoves installed in 9 states of Mexico [29,43,44].

Efficient Charcoal Furnaces

An estimated ten million urban households use charcoal in Mexico. Charcoal is produced mainly in traditional earth kilns with low transformation efficiency of coal burning (only 12% to 20% of the dry weight of the wood is recovered as carbon). This technology cause's harm to the health of producers by inhalation of toxic gases and exposure to high temperatures. It is estimated that domestic consumption of charcoal reaches over 600 thousand ton/year [45]. The improved technologies offer an opportunity to reduce the consumption of wood in the manufacture of charcoal, mitigate greenhouse gas emissions, improve working conditions and the income of farmers, and in general, a more sustainable bioenergy production process.

Brick Furnaces

Brick furnaces were originally designed by the Technological Institute of Minas Gerais, Brazil [47], with theobjective of improving the production process over traditional earth kilns. The transfer and adaptation of technology in Mexico has been the responsibility of Nature and Development Civil Association, an association started in the city of Queretaro in 2003, with a transferred technology known as "RaboQuente". So far eight states are involved, with around 30 projects, 60 furnaces and 100 direct beneficiaries. Main features are shown in Table 3 [46].

Table 3: Projects for use of charcoal furnaces.

Project	Location	Start date	Capacity	Caracteristics
Brick furnaces	Tamaulipas, Jalisco, Queretaro, Hidalgo, Guanajuato, Campeche, Tabasco y Quintana Roo	2003	6 m³ of wood. Productivity: 1,300-2,300 of charcoal, depending on moisture of wood.	Internal diameter: 3.2 m, internal height: 2.20 m, capacity: 6 m³ of wood

Biofuels

Biodiesel is a mixture of fatty acid esters with short chain alcohols, resulting from the reaction of vegetable oils or animal fats with methanol or ethanol at atmospheric pressure [48]. It can completely replace petroleum diesel (B100), or be used in mixtures with different percentages (B1,B5,B10). Its main advantage is that it can be produced from renewable sources [49]. Particularly, biodiesel is drawing attention from government agencies, since it can be a notable factor for promoting regional development in the country [4]. Biodiesel is expected to replace 7.8% of conventional diesel fuel consumption by 2031 [50]. Mexican studies have determined that production of biodiesel can be competitive if the following crops are used as feedstock: oil palm tree, sunflower, soy and Jatrophacurcas [51]. There are different projects being developed in Mexico (Table 4).

Table 4: Projects for biodiesel production.

Project	Location	Start date	Capacity	Source of energy
Bioenergetic Chiapas	Chiapas	2009	2,000 L/d (Tuxtla Gutierrez), 8,000 L/d (Puerto Chiapas), 20,000 L/d (Puerto Chiapas)	Palm oil and Jatrophacurcas
Biocombustibles Internacionales S.A. de C.V.	Nuevo Leon	2005	1.5 million L/d	Animal thallow and recycled oil

Bioenergetic Chiapas

This program was operated by the Institute for Agricultural Restructuring and Tropical Agriculture (IRPAT). It included the establishment of plantations, oil extraction and biodiesel production plants. The objective of the program was to establish 20,000 ha of pinion (Jatrophacurcas L.)

by 2012 in more than 20 municipalities, which was partially achieved. Biodiesel was also obtained from the oil palm and vegetable oils, and it was intended to supply 113 units of the urban public transport of the state capital, Tuxtla Gutierrez and of the city of Tapachula. These vehicles used B5 and B20, but there was a unit that used B100. In April 2010, two biodiesel plants were installed. One in Tuxtla Gutierrez, with Swedish technology, which produces 2000 L/d; another in Puerto Chiapas, with a module of Colombian- Mexican technology (whose production is 8000 L/d) and other English technology module (which produced 20 000 L/d). Overall, the installed production capacity was 30,000 L/d [29,50].

BiocombustiblesInternacionales S.A. de C.V

The municipality of Cadereyta, Nuevo León holds the first biodiesel production plant in Mexico, with a capacity of 1.5 million liters a month. The main raw materials used are beef tallow and recycled vegetable oils. It began operations in October 2004 and was officially opened in July 2005. The produced biodiesel at the plant is used by PEMEX Refining as an additive of ultra-low sulfur diesel due to its high capacity as a lubricant because it contains no sulfur. The introduction of biodiesel in Mexico would bring out several benefits, such as: job creation, expansion of the facility to rural areas, development of rural economies, conservation of oil resources and development of multiple crops [4,52].

CONCLUSIONS

Energy policy in Mexico has strengthened the development of renewable sources of energy during the last years; however, roadmaps to define its participation in the energy mix are still missing. One of the most promising sources of renewable energy is biomass, which has an estimated potential of 3,569 PJ/year, where almost 50% would come from the management of native forests. Biomass conversion technologies in Mexico rely mainly on biogas and biofuel production, wood stoves and charcoal furnaces which have representative projects in various states of the country. Three important wood stove types (Patsari, Mexalit and Onil) and one charcoal furnace (RaboQuente) were developed in Mexico. The country has been identified as a pioneer in the landfill biogas energy generation in Latino America. Moreover, the government is mindful of the production of biodiesel to promote regional development in the country, especially in rural communities. Development of this kind of projects will allow the country to reduce gradually its dependence on fossil fuels.

ACKNOWLEDGEMENTS

The authors thank the Chair of Environmental Bioprocesses (Tecnológico de Monterrey) for the financial support given during this investigation, to Red Mexicana de Bioenergía A.C. for its great effort to collect information and finally to SolucionesTecnológicas con Microalgas S.A. de C.V. for its support during this study.

REFERENCES

1. AlemÃ¡n-Nava GS, Casiano-Flores VH, CÃ¡rdenas-ChÃ¡vez DL, DÃaz-Chavez R,Scarlat N, et al. (2014) Renewable energy research progress in Mexico: Areview. Renew Sustain Energy Rev, 32:140-153.

2. REEEP (2007)Sustainalbe energy policy initiative for Latin Amertrica and the Caribbeanreport.

3. (DOF). DO de la F. Ley de promociÃ³n y desarrollo de los bioenergÃ©ticos; 2008 n.d.

4. Rangel-HernÃ¡ndez VH, GÃ³mez-Vargas DP, Gallegos-MuÃ±oz A,Plascencia- Mora H (2011) The development impact of biodiesel: a review of biodieselproduction in mexico. Int J Energy Environ Eng, 2:91-99.

5. Camara de Diputados. Law for the Use of Renewable Energy and Finance of the Energy Transition. D Of La Fed 2008:1-12.

6. SENER. Anhydrous Ethanol Program Introduction 2011.

7. Network R (2013) Renewables 2013-Global Status Report.

8. SENER (2011) National Energy Balance.

9. Masera O. (1993) Sustainable fuelwood use in rural Mexico. Patterns Resour Use, USALawrence Berkeley.

10. INEGI (2009)EstadÃsticashistÃ³ricas de MÃ©xico.

11. The World Bank (2014) World Development Indicators: Rural environment and land use.

12. Castillo-Santiago MÃ, Ghilardi A, Oyama K, HernÃ¡ndez-Stefanoni JL, Torres I, et al. (2013) Estimating the spatial distribution of woody biomass suitable for charcoal making from remote sensing and geostatistics in central Mexico. Energy Sustain Dev,17:177-88.

13. GarcÃa-Frapolli E, Schilmann A, Berrueta VM, Riojas-RodrÃguez H, Edwards RD,et al. (2010)Beyondfuelwood savings:Valuing the economic benefits of introducing improved biomass cookstoves in the PurÃ©pecha region of Mexico. EcolEcon 69:2598-2605.

14. Toscano L, Montero G, Stoytcheva M, Campbell H, Lambert A (2011) Preliminaryassessment of biodiesel generation from meat industry residues in Baja California,Mexico. Biomass and Bioenergy, 35:26-31.

15. Sheinbaum-Pardo C, CalderÃ³n-Irazoque A, RamÃrez-SuÃ¡rez M (2013) Potential of biodieselfrom waste cooking oil in Mexico. Biomass and Bioenergy, 56:230-238.

16. MurilloAlvarado PE, PonceOrtega JM, CastroMontoya AJ, SernaGonzÃ¡lez M,El-Halwagi MM (2014) 24th European Symposium on Computer Aided ProcessEngineering. Elsevier.

17. Aguilar Virgen Q, TaboadaGonzÃ¡lez P, OjedaBenÃtez S, CruzSotelo S (2014) Powergeneration with biogas from municipal solid waste: Prediction of gas generation within situ parameters. Renew Sustain Energy Rev 30:412-419.

18. Aguilar Virgen Q, TaboadaGonzÃ¡lez P, OjedaBenÃtez S (2014) Analysis of thefeasibilityof the recovery of landfill gas: a case study of Mexico. J Clean Prod, 79:53-60.

19. HalsnÃ¦s K (1996)The economics of climate change mitigation in developing countries.Energy Policy 24:917-26.

20. Pine K, Edwards R, Masera O, Schilmann A, MarrÃ³nMares A, RiojasRH (2011)Adoption and use of improved biomass stoves in Rural Mexico. Energy Sustain Dev15:176-83.

21. Holmes HA, Pardyjak ER, Speckart SO, Alexander D (2011) Comparison ofindoor/outdoor carbon content and time resolved

PM concentrations for gas andbiomass cooking fuels in Nogales, Sonora, Mexico. Atmos Environ 45:7600-11.

22. Saatkamp BD, Masera OR, Kammen DM (2000) Energy and health transitions indevelopment: fuel use, stove technology, and morbidity in JarÃ¡cuaro, MÃ©xico.Energy Sustain Dev, 4:7-16.

23. Aldana H, Lozano FJ, Acevedo J (2014) Evaluating the potential for producing energyfrom agricultural residues in MÃ©xico using MILP optimization. Biomass andBioenergy, 67:372-389.

24. Energy Information Administration. International energy statistics:Mexico n.d. U.S.A.

25. Islas J, Manzini F, Masera O (2007)A prospective study of bioenergy use in Mexico.Energy 32:2306-2320.

26. Valdez Vazquez I, AcevedoBenÃtez JA, Santiago CH (2010) Distribution andpotential of bioenergy resources from agricultural activities in Mexico. RenewSustain Energy Rev,14:2147-2153.

27. ECLAC (2008) Statistical Yearbook for Latin America and theCaribbean.

28. ProMÃ©xico. EnergÃasRenovables (2013). Secr Econ.

29. Cerutti OM (2010) La bioenergia en Mexico, casos de estudio.

30. Johnson T, Alatorre C, Zayra R (2009) MÃ©xico: estudiosobre la disminuciÃ³n de emisionesde carbono.

31. REMBIO (2011) Bioenergy in Mexico, current situation and outlook.

32. McKendry P (2002) Energy production from biomass (part 2): conversion technologies.BioresourTechnol 83:47-54.

33. Ryckebosch E, BermÃºdez SPC, TermoteVerhalle R, Bruneel C, Muylaert K, et al. (2013) Influence of extraction solvent system on the extractability of lipidcomponents from the biomass of Nannochloropsisgaditana. J ApplPhycol26:1501-1510.

34. Cuellar-Bermudez SP, Garcia-Perez JS, Rittmann BE, Parra-Saldivar R (2014)Photosynthetic bioenergy utilizing CO2: an approach on flue gases utilization forthird generation biofuels. J Clean Prod.

35. LFGConsult (2007) Case of CDM Landfill Gas Projects Monterrey, Mexico (BENLESA).

36. AGCert (2005) AWMS GHG Mitigation Project MX05-B-09 , Nuevo LeÃ³n , MÃ©xico.

37. Global Methane Initiative (2010) Resource Assessment for Livestock and Agro-Industrial Wastes â€"Mexico.

38. Curry N, Pillay P (2012) Biogas prediction and design of a food waste to energy system forthe urban environment. Renew Energy, 41:200-209.

39. Masera O (2010)Patrones de uso de leÃ±a en MÃ©xico: situaciÃ³n actual y perspectivas a largoplazo.

40. Fonseca HL (2011)EmisionesporConsumoDomÃ©stico de LeÃ±a.

41. Masera O, Edwards R, Arnez CA, Berrueta V, Johnson M, et al. (2007) Impactof Patsari improved cookstoves on indoor air quality in MichoacÃ¡n, Mexico. EnergySustain Dev 11:45-56.

42. Ruiz-Mercado I, Masera O, Zamora H, Smith KR (2011) Adoption and sustained use ofimproved cookstoves. Energy Policy, 39:7557-66.

43. Adkins E, Chen J, Winiecki J, Koinei P, Modi V (2010) Testing institutional biomasscookstoves in rural Kenyan schools for the Millennium Villages Project. EnergySustain Dev 14:186-93.

44. Tyagi SK (2013) Design, development and technological advancement in the biomasscookstoves: A review. Renew Sustain Energy Rev, 26:265-85.

45. Bravo R, Cerutti O, Chalico T (2010)Woodfuels and climate change mitigation. Casestudies from Brazil, India and Mexico. Clim Chang.

46. Oliveira AC, Salles TT, Pereira BLC, Carneiro A de CO, Braga CS, et al (2014) Economic viability of charcoal production in two production systems. Floresta,44:143-52.

47. Monteiro M (2000) Metallurgy in the Brazilian Amazon: Alternatives for activities withscarce ecological prudence. Pap Do NAEA, 157:1-28.

48. Chisti Y (2007) Biodiesel from microalgae. BiotechnolAdv, 25:294-306.

49. Demirbas A (2008) Biofuels sources, biofuel policy, biofuel economy and global biofuelprojections. Energy Convers Manag, 49:2106- 16.

50. Lozada I, Islas J, Grande G (2010) Environmental and economic feasibility of palm oilbiodiesel in the Mexican transportation sector. Renew Sustain Energy Rev,14:486-92.

51. SENER S de E (2006) Potentials and Feasibility use of the Bioethanol and Biodiesel in theMexican Transport. SecrEnerg.

52. Masera O, RodrÃguezMartÃnez N (2006)Potenciales y viabilidaddeluso de bioetanol ybiodiesel para el transporte en MÃ©xico.

Chapter 5

Enhancement of Crude Oil Biodegradation by Immobilizing of Different Bacterial Strains on Porous Pva Hydrogels or Combining of Them With their Produced Biosurfactants

Gibrán S Alemán-Nava[1], Luisaldo Sandate-Flores[1], Alexander Meneses-Jácome[2], Rocío Díaz-Chavez[3], Jean-Francois Dallemand[4] and Roberto Parra[1]*

Abdeen Z[1]*, Huda K El-Sheshtawy[2] and Moustafa YMM[3]

[1]Petrochemical Department, Egyptian Petroleum Research Institute, Nasr City, Cairo, Egypt

[2]Processes Development Department, Egyptian Petroleum Research Institute, Nasr City, Cairo, Egypt

[3]Evaluations and Analysis Department, Egyptian Petroleum Research Institute, Nasr City, Cairo, Egypt

ABSTRACT

The degradation of the crude oil in wastewater by each of freely microorganisms and by immobilized them on each of crosslinked poly (vinyl alcohol) hydrogel (CPVA) and its foam (CPVAF) was reported. Also, it was studied by using biosurfactants (Bios) with free cells. The macroporous CPVAF was prepared by adding $CaCO_3$ as poreforming agent and epichlorohydrin as crosslinker. The prepared polymers are examining by FTIR, XRD, TGA, DSC and SEM analysis. The microorganisms of B.l., R.e. and P.x. isolated from contaminated effluents were investigated. The ability of these microorganisms to degrade the n-paraffin and PAHs was assessed by GC and HPLC analysis, respectively. Moreover, their stabilities and activities were tested in the growth count of bacteria study. The crosslinked CPVA carrier demonstrated better thermal stability and improvement in the microorganism efficiency with respect to hydrocarbons degradation than these of the CPVAF carrier. Scanning electron microscopy showed the presence of extracellular structures that could play an important role in the immobilization stability of cells on polymers. As well, GC analysis revealed that the percentage biodegradation ability of immobilized cells R.e. on CPVAF for the total n-paraffin was approximately, 100%. While, the HPLC analysis showed that the percentage biodegradation of cells for PAHs was enhancement by immobilized them on CPVA and also, at adding Bios to them. The results suggest that the potential of using each of CPVA, PVAF as cell carriers and Bios separately, to free cells to enhancing the biodegradation of petroleum hydrocarbons in an open marine environment.

INTRODUCTION

Biodegradation of petroleum hydrocarbons is a complex process that depends on the nature and the amount of the hydrocarbons present, saturates, aromatics, asphaltenes and the polar compounds [1]. The polycyclic aromatic hydrocarbons (PAHs) [2] are a minor constituent of crude oils; however, they are widespread environmental pollutants. They are generated from petroleum and many pyrolysis processes. They form concern because of their potentially deleterious effects on human health and many can be recalcitrant in the environment

[3]. Bioremediation can be described as the conversion of chemical compounds by living organisms, especially microorganisms, into energy, cell mass and biological waste products [4]. Bacteria can convert PAHs completely to biomass, CO_2, and H_2O. A wide range of different bacteria is able to completely assimilate a defined range of compound, or exhibit just partial metabolism [5]. Therefore, the use of pure cultures of microorganisms, specially adapted to metabolize the contaminant, is envisaged as an attractive alternative. Rhodococcus erythropolis is well known microorganism containing a large set of enzymes that allows carrying out an enormous number of bioconversion and degradation [6,7]. Biosurfactants are biological compounds that exhibit high surface-active properties [8] and are produced by a wide variety of microbes. Wide spectra of microbial compounds, including glycolipids, lipopeptides, fatty acids, and polymeric biosurfactants, have been found to have surface activity [9]. They have important advantages, such as biodegradability, low toxicity, and various possible structures, relative to chemically synthesized surfactants [10]. So, they were use in environmental applications, such as in bioremediation [11] and in enhanced oil recovery [12]. Microbial cells can be immobilized on various hydrophilic polymeric entrapment matrices such as hydrogels [13].

Among hydrophilic polymers, poly (vinyl alcohol) PVA. poly (hydroxyethylmetacrylate) PHEMA and others, are the most often used for hydrogel's synthesis [14-17]. Poly (vinyl alcohol) (PVA) is a water-soluble material that has been widely used for immobilization of bioactive materials [18]. Because of, PVA is innocuous for bioactive matter and possesses many attractive properties [19], it is a potential bio-carrier material that can be applied in the fermentation industry [20], chemistry [21], and the environment [22]. The chemical stability and mechanical properties can be improved by physical or chemical crosslinking [19]. Nevertheless, research on the preparation and use of PVA foam for microbial immobilization is rarely reported. Therefore, there is a great scope for designing better PVA foam carriers for immobilizing microorganisms and developing new technologies for wastewater bio-treatment. In this work, hydrogel composites based on PVA and calcium carbonate were prepared and foamed by adding hydrochloric acid, to produce a macroporous carrier [23]. Epichlorohydrin was then used as a chemical crosslinking agent to form network structures as well as to improve the foam's stability. Further investigation has revealed the desirable properties of crosslinked PVA foam (CPVAF) as a carrier of immobilized microorganisms. A crosslinked prepared PVA (CPVA) carrier was chosen for comparison and it was used for the preparation of the immobilization matrix. The main advantages in the use of immobilized cells in comparison with suspended ones include the retention in the reactor of higher concentration of microorganisms, easy removal of bacteria after use from the reaction mixture, providing the ability to control reaction time, reuse of cells for many reaction cycles, lowering the total production cost of cells-mediated reactions, provide pure products [24-26]. The efficiencies of each of free, immobilized cells *(Bacillus licheniformis, Rhodococcuserythropolishas and Pseudomonas xanthomarina)* on CPVAF and CPVA matrixes crosslinked by epichlorohydrin [17], and their biosurfactants in degradation of crude oil in water were evaluated. In addition, we evaluated the capability of each of CPVA and CPVAF in water, as immobilizers to the enhancement the degradation efficacy of each microorganism to degrade the pollutants.

MATERIALS AND METHODS

Materials and Microorganism

The PVA used in this study was analytical grade and was purchased from Merck, Germany. The average Mw of the PVA was 127,000 and the degree of hydrolysis was 89%. Epichlorohydrin (EP), calcium carbonate, sodium hydroxide, hydrochloric acid and other chemicals were all obtained from Beijing Chemical Reagent Factory (China) and used without further purification. The crude oil used in the present work, was Asphaltenic crude oil. The microorganisms used in this study were, *Bacillus licheniformis* ATCC10716 (B.l.) and *Rhodococcus erythropolis* ATCC13260 (R.e.) supplied by the microbial resources center (MIRCEN), Faculty of Agriculture, Ain Shams University, Cairo, Egypt. Also, *Pseudomonas xanthomarina* KMM 1447 (P.x.) bacterial was isolated from Gemsa Bay (Red Sea).

Molecular Identification of the Bacterial Strain Isolated from the Red Sea

An analysis of 16S rRNA was performed to taxonomically character-ize the isolated strains (Sigma Scientific Services Co., Egypt). The cell of the bacterial strain was harvested through the enrichment medium up to 2×10^9 bacterial cells. DNA was extracted using protocol of Gene Jet genomic DNA purification Kit (Thermo) (Sigma Scientific Services Co., Egypt). To amplify the 16S rDNA gene, a polymerase chain reaction (PCR) was performed using two primers, the forward primer (5′-AGA GTT TGA TCC TGG CTCA-3′) and the reverse primer (5′-GGT TAC CTT GTT ACG ACT-3′). PCR was cleaned up to the PCR product using GeneJET™ PCR Purification Kit. A 45 ul of Binding Buffer was added to completed PCR mixture. The mixture was then thoroughly, transferred from step 1 to the GeneJET™ purification column. The mixture was centrifugated for 30-60 s at >12000 x g, then the flow were discarded. A 100 ul of wash buffer was added to the GeneJET™ purification col-umn and centrifuged for 30-60 s. Then, discarded the flow-through and placed the purification column back into the collection tube. The mixture was centrifugated the empty GeneJET™ purification column for an additional one min. to completely remove any residual wash buffer. The purification column was transferred to a clean 1.5 ml mi-cro centrifuge tube. A 25 ul of elution buffer was added to the center of the GeneJET™ purification column membrane and centrifuged for one min. The GeneJET™ purification column was discarded and the purified DNA was stored at -20°C. Following purification of the PCR products, the DNA sequence of the positive clone was subjected to a similarity search BLAST on the NCBI website (http://www.ncbi.nlm. nih.gov), and deposited into GenBank. Many relevant 16S rRNA gene sequences with validly published names were selected as references from the Gen- Bank.

Culture Conditions

Culture medium and production of different biosurfactants: The bacterial strains were streaked on a nutrient agar slants and incubated for 24 h at 30°C. Two loops of each culture were inoculated in 40 ml of nutrient broth in a 100 ml Erlenmeyer flask and incubated in a rotary shaker 150 rpm at 30°C for 8-12 h until cell numbers reached 108 CFU/ml. For biosurfactant synthesis, a mineral salt medium with the following composition was utilized: 2.0 g/l of Na_2HPO_4, 2.0 g/l of KH_2PO_4, 0.01 g/l of $MgSO_4.7H_2O$, 2.5 g/l of $NaNO_3$, 0.8 g/l of NaCl, 0.2 g/l of $CaCl_2$, 0.8 g/l of KCl, 0.001 g/l of $FeSO_4$. $7H_2O$ and 5 ml of a trace element solution. Trace element solution contained 0.116 g/l of $FeSO_4\cdot7H_2O$, 0.232 g/l of H_3BO_3, 0.41 g/l of $CoCl_2\cdot6H_2O$, 0.008 g/l of $CuSO_4\cdot5H_2O$, 0.008 g/l of $MnSO_4\cdot H_2O$, 0.022 g/l of $[NH4]6Mo7O24$ and 0.174 g/l of $ZnSO_4$ [27]. The carbohydrate (glucose) was added to make a final concentration 2%, and the concentration of yeast extract was 3%. Cultivation studies have been done in 500 ml flasks containing 150 ml medium at 30°C for 72 h.

Preparation of PVA Hydrogel Crosslinked by Epichlorohydrin (CPVA) [16]

CPVA was prepared by adding PVA to 100 ml of distilled water and blending by a mechanical stirrer in a boiling water bath for 1 h. After the mixture cooled to room temperature, stoichiometric amounts of each of epichlorohydrin (EP) as the crosslinker and potassium hydroxide solution were added separately to the mixture with stirring it gently, at 35°C for 2 h. The amount ratios of PVA/E/KOH were: 100/20/25, 100/60/65, 100/100/105 and 100/125/130, 100/140/150. Five CPVA hydrogel disc samples were prepared.

Preparation of Crosslinked PVA Hydrogel Foam (CPVAF)

CPVAF was prepared as follows: calcium carbonate powder and PVA were added in ratio 1:2 to 100 ml distilled water and blended by a mechanical stirrer in a boiling water bath for 1 h. After the mixture cooled to room temperature, the certain amount of hydrochloric acid (5 M) was added and the mixture was stirred vigorously to dissolve the calcium carbonate and produce CO_2 gas bubbles. Then the molded foam was submerged in 200 ml NaOH (1 M) containing 1 ml of EP and stirred gently at 35°C for 2 h. All prepared CPVA and CPVAF gels were cut into discs of about 0.5 cm thickness and 1.2 cm diameter and maintained in sterile conditions until their use for cell immobilization [24].

Biodegradation Test of the Crude Oil Using Free Cells in Presence/ Absence of Different Biosurfactants and by Immobilizing of these Cells on (CPVA and CPVAF) Gels

To assess the efficiency of different bacterial strains to degrade the crude oil, 2 ml of inoculum (7.5 log count) was inoculated into the medium mineral salt medium MSM (100 ml) in a 250 ml erlenmeyer flask. The cultures were incubated in a temperature controlled shaker incubator at 150 rpm at 30°C for 15 days, using (1g) crude petroleum oil as a sole carbon source with/ without adding the discs of CPVA and CPVAF gels separately. The biosurfactant was produced by using the medium MSM with (1g) glucose as a sole carbon source for 72 h followed by addition of (1 g) crude oil until the end of incubation period. A sample without inoculum was taken as a control. Hence, the flasks were incubated at 30°C, 150 rpm, pH 7.5 for 15 days [28]. After incubation period, the bacterial count was determined and the crude oil samples were extracted from different microcosm and gravimetric analysis was also determined as follow:

Determination of the Growth Counts for Different Bacterial in Presence/ Absence Biosurfactant and at Immobilized these Bacterial on CPVA and CPVAF Gels: The bacterial counts were implemented by using the plate count technique where the different microcosms samples (1ml) was serially diluted in a sterile saline. Then inoculation in Luria broth (LB) plates medium containing (g/l of distilled water), 10.0 g/l of NaCl, 10.0 g/l of tryptone and 5.0 g/l of extracted yeast, the medium was adjusted to pH 7.0. The cultures were then incubated at 30°C for 48 h, then; the plate count in the range 30-300 Calonies was recorded. The bacterial count /ml (average colony forming unit (CFU)/ ml) was calculated from the following equation:

Bacterial count / ml = colony count / plate x dilution factor [29].

Gravimetric Estimation for the Extracted Crude Oil after its Biodegradation: After incubating period (15 days), the polluted bacterial broth (100 ml) was thoroughly shaken with carbon tetrachloride (50 ml 3 times) in a separating funnel and the three fractions were collected in case of crude oil samples. The collected organic layer was dried over anhydrous sodium sulphate and the solvent was removed by a rotary evaporator till reached a constant weight. The oil sample was accurately weighed, the percentage of the biodegradation oil was calculated and the alteration in its chemical composition was studied by chromatographic analysis (GC and HPLC) [2,30].

Characterization of CPVA and CPVAF Hydrogels

The structures of carriers CPVA and CPVAF were characterized before and after foaming, by using the FT-IR spectra that, recorded using a Nicolet IS-10 FT-IR spectrophotometer. Also, the thermal stabilities of the CPVA and CPVAF carriers were studied using a Thermogravimetric analyzer (TGA) and Differential scanning calorimetry (DSC). All TGA spectra were recorded under a nitrogen atmosphere up to 600°C using a programmed rate of 10°C/min and the DSC spectra were obtained with a DSC-30 under N2 atmosphere at a heating rate of 20°C/min. X-ray diffraction patterns were recorded on X` Pert (Berlin, Germany) D500 diffract meter with a back mon°Chromatic and a Cu anticathode. The phase morphology was studied using a JSM-T20 (JEOL, Tokyo, Japan) scanning electron microscope (SEM). For scanning electron observations, the surface of the sample was mounted on a standard specimen stub. A thin coating (~ 10-6 m) of gold was deposited into the sample surface and attached to the stub prior to SEM examination in the microscope to avoid electrostatic charging during examination. The Swelling behavior of the prepared hydrogels was estimated by immersing a known weight of the sample disc in solutions of pH 7 at 30°C until the swelling equilibrium was reached. The disc was removed, dried with absorbent paper to get rid of excess water then weighed. The degrees of swelling (DS) for these samples were calculated at time intervals (5, 15, 30, 60, 120, 180 minutes) with respect to CPVA (A) and CPVAF (B) samples, with the following equation:

DS = (m- m') / m' (1),

Where m and m' denote the weights of hydrogel and dried hydrogel sample, respectively [31].

Gas Chromatographic Analysis

Biodegradation of the crude oil was monitored using an Agilent 6890+ gas chromatography instrument according to the testing method IP 318 [32]. The detector was a flame ionization detector 325°C, separation was completed on HP-1 capillary column (100% methyl silicone siloxane, 30 m length, 0.35mm internal diameter and 0.25 mm thickness film), a dose of 0.5 µl was used using a splitter injector 300°C and the oven temperature 100-320°C reached applying a heating rate of 5°C/min. The carrier gas was nitrogen (2ml/min). Identification of the n-paraffins peaks was confirmed by chromatography a reference mixture of n-paraffins of a known composition under the same operating conditions.

High Performance Liquid Chromatographic (HPLC) Analysis [33]

The Identification and quantification of PAHs in the crude oil remained after biodegradation and the corresponding control sample were performed using HPLC. The apparatus used was model Waters HPLC 600 E, equipped with dual UV absorbance detector, model Waters 2487 and auto-sample Waters 717 plus attached to computerized system with millennium 3.2 software. The condition of separation [34] is as follow: Column (Supelcosil.LC-PAH, 5 µm particles, 15 cm length and 4.6 mm internal diameter), Mobile phase (gradient acetonitrile: water 60 to 100% acetonitrile (v/v) over 45 minutes), Flow rate (0-2 min. 0.2 ml/min, 2-45 min. 1.0 ml/min) and Dector (Set at 254 nm).

RESULTS AND DISCUSSIONS

Hydrogels Characterization

PVA contributed to strength and durability of the carriers. To get high strength, the percentage of PVA in the experiment was 10% concentrations, the higher than 10% resulted viscous solutions [15]. The amounts and the ratio of the pore-forming agents are the important factors which decide the aperture uniformity and size distribution, and the elasticity of the foams. The infrared spectra of PVA crosslinked (CPVA) [16] and crosslinked PVA foam (CPVAF) are represented in Figure 1A and 1B, respectively.

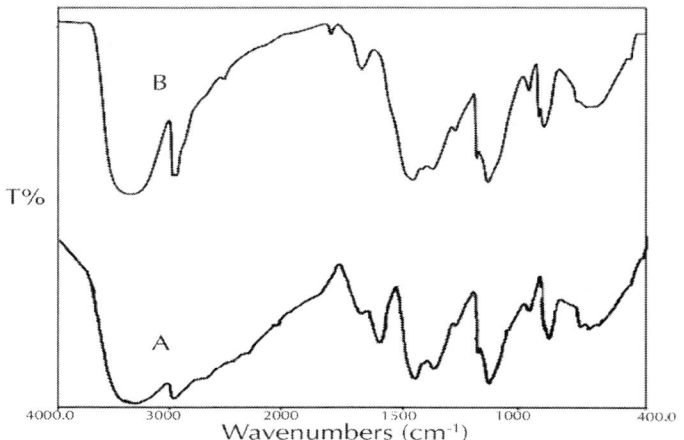

Figure 1: FT-IR spectra of CPVA (A) and CPVAF (B) carriers.

FTIR spectra shows broad peak at 3440 to 3100 cm⁻¹ reveals stretching vibrations of the hydroxyl groups and the peak at 2940 to 2900 cm⁻¹ is due to C-H stretching vibrations. The CPVA spectra shows abroad bonded hydroxyl peak around 3263 cm-1, but the hydroxyl peak of CPVAF was obviously less condensed broad, and was shifted to 3313 cm⁻¹. The intramolecular and intermolecular hydrogen bonds of the OH groups of PVA and other molecule, shifted the band of –OH group to lower frequencies as shown in spectra of CPVA .The $CaCO_3$ shows absorption bands at 1419, 871and 713 cm⁻¹, which are attributed to the Ca-O Stretching vibration and bending vibration. This result shows that network structures were formed on the chains of the CPVA with a certain amount of hydroxyl. The stronger peak of CPVA and CPVAF carriers were at 1090 cm⁻¹ (C–O–C), as a result of the formation of crosslinked network structures. FTIR spectra showed the bands at 1646, 1568 and 1654 cm⁻¹ indicated presence of the residual vinyl acetate groups in the PVA chains (C=C) and the traces of water molecules (bending vibration). These peaks are very weak in spectra of CPVAF due to the more crosslinked hydrogels, than in the CPVA one. The peak around 1095 cm^{-1} denotes C-O stretching of the secondary alcoholic groups [16]. The swelling behaviors of the chemically crosslinked CPVA [20] and CPVAF carriers, pH (7) at 30oC were represented in Figures 2 and 3.

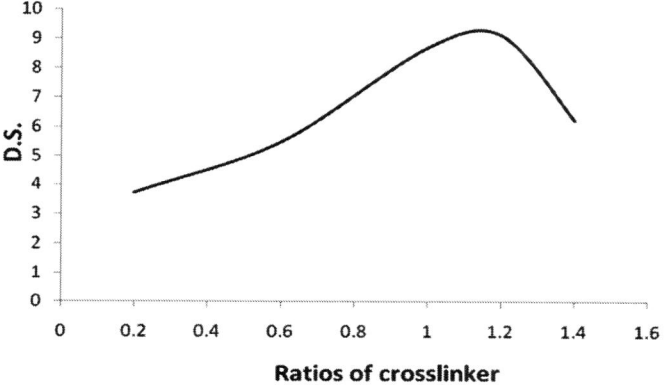

Figure 2: The degree of swelling (D.S) of CPVA at different ratios cf cross-linker.

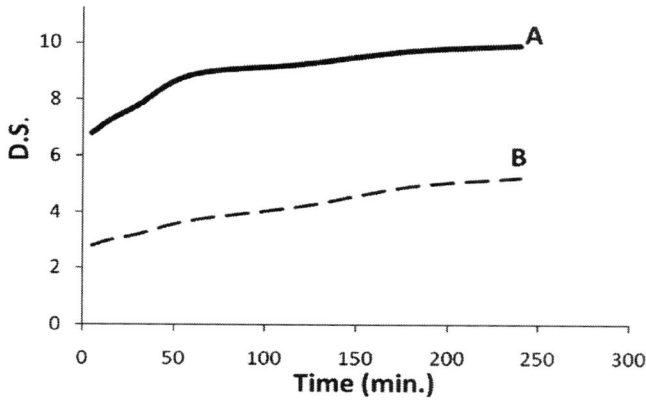

Figure 3: Swelling ratio of CPVA (A) and CPVAF (B) as a function of incubation time at pH =7.

The equilibrium swelling rates versus concentrations or the crosslinker ratios relationship is represented in Figure 2. The water uptake is found to decrease for CPVA matrix at increasing the ratio of crosslinker than certain ratio in PVA/E/ KOH (100/125/130). This behavior can be explained by the existence of an important crosslinking density which is in accordance with an increase in the crosslinker ratio to optimum swelling properties (9.1), after what , this property decrease (from 9.1 to 6.2) with an increasing in crosslinking density [20]. For CPVA and CPVAF, water uptake increases with increase of time, but the swelling property of CPVA is higher than in CPVAF and this was may suggested for occurring a high crosslinking density in CPVAF resulting in lowering the water uptake. Also, The PVAF hydrogel has the more porous with large size that may be decrease the retention water and, furthermore the presence the oxygen of $CaCO_3$ may form physical bonding with hydroxyl group of PVA leading to a decrease in the penetration of water molecules into the hydrogel network. The thermal analysis of the carriers prepared was used to compare the differences in the polymeric arrangements of the carriers. The thermogravimetry (TG) curves are presented in Figure 4.

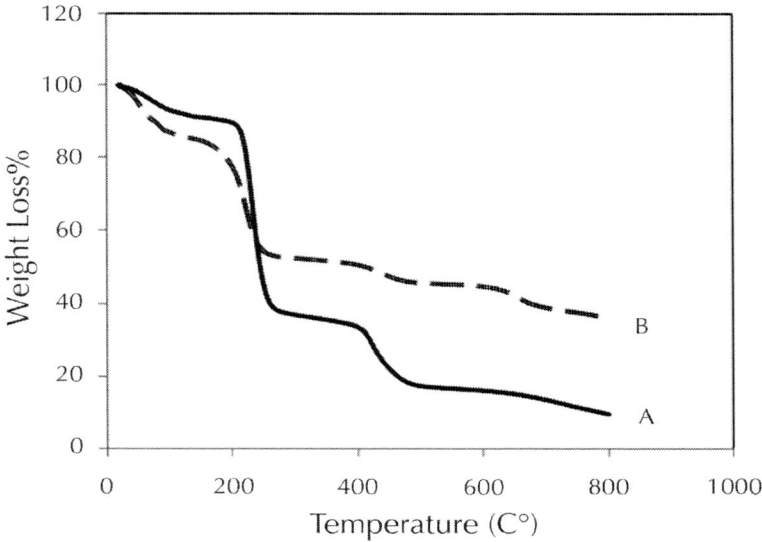

Figure 4: TGA curves of CPVA (A) and CPVA F (B) carriers.

The CPVA carrier had three main degradation steps: the first was at 189 and 260°C and 10% loss weight was ascribed to the release of volatile compounds, mainly water, which were present in the hydrophilic material. The second stage was between 260 and 400°C regarded as the elimination of acetic acid to form a polyene, whereas the third step was between 400 and 485°C corresponding to the breakage of the main chain. There was total 82% lost at last might be attributed to the decomposition of CPVA. TG curves for the CPVAF carrier also showed three consequent stages: The first was come between 160 and 270°C, and the weight loss was increased to 14% at 160°C, indicated an increase in carrier hydrophilicity than in CPVA [3]. The second stage was from 270 to 578.0°C and finally, the last stage was from 578.0 and 790.8°C. There was total 64% lost at last temperature. Also, the degradation events showed an increase in the thermal stability of the carrier as a result of adding foaming reagent to crosslinking PVA. The DSC curves of the carriers are presented in Figure 5.

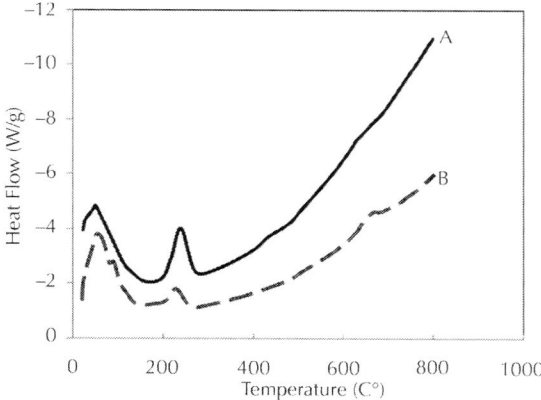

Figure 5: DSC curves of CPVA (A) and CPVA F (B) carriers.

The sample carriers were included to establish a detailed behavior of the organization of polymeric chains. CPVAF exhibited an endothermic peak at about 223°C, corresponding to the melting temperature of PVA. The melting temperature was much increased for CPVA. This implied that foaming limited the mobility of the PVA chains. The addition of $CaCO_3$ to the polymeric carrier increased the distance between the chains, regardless of whether the polymer became more crosslinked or branched, which made the organization of PVA in crystalline lattices difficult and formed network structures. The XRD patterns for CPVA (A) and CPVAF (B) carriers were given in Figure 6.

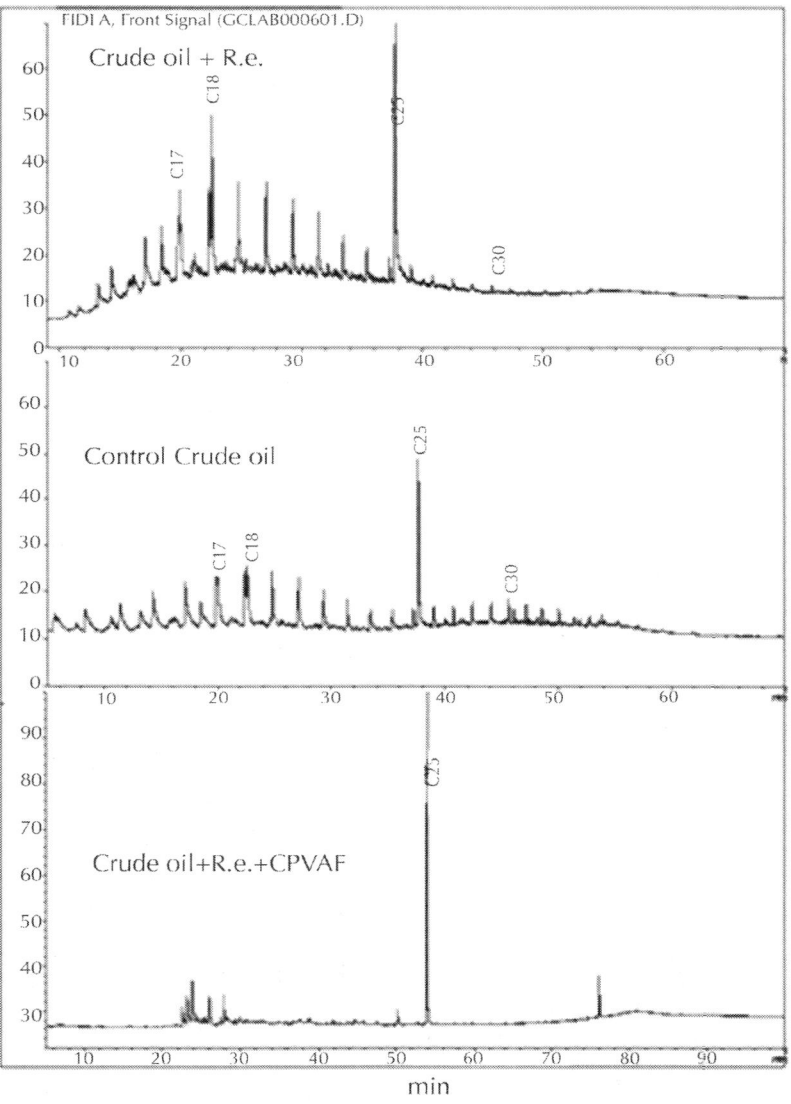

Figure 6: XRD patterns of CPVA (A) and CPVA F (B) carriers.

The diffraction peak at around 18.4°C was a characteristic peak of the crystalline structure of PVA and it was shifted to less 2θ due to the some deformation in PVA crystallinty as a result of crosslinking. However, its intensity weakened significantly, indicating that the crystalline structure of PVA was changed after the introduction of $CaCO_2$. It may be due to the intermolecular and intramolecular hydrogen bonding between the carbonyl groups in $CaCO_3$ and hydroxyl groups in PVA, which inhibited the formation of crystal structure. This promotes the formation of physically crosslinked PVA hydrogel in CPVAF so it becomes more crosslinked structure.

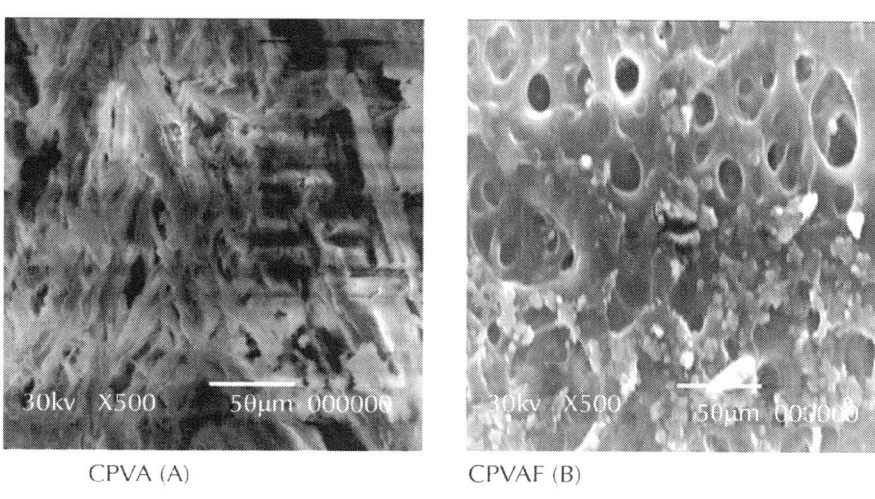

CPVA (A)   CPVAF (B)

Figure 7: SEM photographs of CPVA (A) and CPVA F (B) carriers.

Figure 7 shows the surface morphologies of CPVA (A) and CPVAF (B) by scanning electron microscope. It can be seen from Figure 7B that CPVAF were perfectly porous in shape with a rough surface. But, Figure 7A of CPVA shows the less pour in size than in 7B of CPVAF, also, it can be seen that the obtained surfaces of each CPVA (A) and CPVAF(B) were porous and the pore sizes were heterogeneous because reaction between $CaCO_3$ and HCl was not uniform throughout the samples. So, CPVA has the more narrow pours, as a result it has the properties are favorable for producing good adsorption capacity. Therefore, an ideal hydrogel carrier should possess a highly porous structure to facilitate the non-hindered diffusion of solutes and dissolved gases.

The Chemical Structure of the Biosurfactants Produced by Different Bacterial Strains

Biosurfactants [9,10] are surfactants that are produced extracellularly or as part of the cell membrane by bacteria, yeasts and fungi. They can be potentially as effective with some distinct advantages over the highly used synthetic surfactants including high specificity, biodegradability and biocompatibility. In this work the scope was on certain types of the biosurfactants were produced by different bacterial strains that obtained from contaminated area in our country. The function of these biosurfactant was the increased the solubility of some hydrocarbon compounds so, it enhance the efficiency of microorganisms to degrade a crude oil. It was found that, the biosurfactants are grouped as glycolipids, lipopeptides, phospholipids, fatty acids, neutral lipids, polymeric and particulate compounds [35]. Most of these compounds are either anionic or neutral. Only a few are cationic such as those containing amine groups. The hydrophobic part of the molecule is based on long-chain fatty acids, hydroxy fatty acids or α-alkyl-β-hydroxy fatty acids. The hydrophilic portion can be a carbohydrate, amino acid, cyclic peptide, phosphate, carboxylic acid or alcohol. In this study, the microorganisms (Bacillus licheniformis, Rhodococcus erythropolis *and Pseudomonas xanthomarina*) produced three types of the biosurfactants with different structures, cyclic lipopeptide surfactin, phosphatidyl ethanolamine and rhamnolipids respectivly. The structure of the producing biosurfactant was affected by some factors such as type of microorganism, pH, nutrient composition, substrate and temperature used. Where, in our work we choose the conditions that are suitable for the medium of microorganism in our country.

1. The bacterium Bacillus licheniformis has been produced surfactin biosurfactant with a cyclic structure that has affect on enhanced the ability of microorganism to degrade some of hydrocarbon compounds. Where, it increases the solubility of these compounds to increase the capability of microorganism to degrade them.

2. The bacterium *Rhodococcus erythropolis* has been produced *Phosphatidyl ethanolamine biosurfactant*, with cationic structure contained the amino group with positive charges facility the solubility of some of hydrocarbon compounds and their degraded by cells.

3. The isolated bacterial strain *Pseudomonas xanthomarina* has been produced rhamnolipids. From the literature review the*Pseudomonas* sp. produced rhamnolipids biosurfactant rhamnolipids. This strain was first identified using molecular identification and performed by amplifying and sequencing the 16S rRNA gene sequences. The results of the identification procedure showed that the isolated bacteria belong to the *Pseudomonas xanthomarina* KMM 1447. It was found this type of surfactin has hydrophilic groups (COOH- and OH-) was lower the surface tension and so, the interfacial tensions

at pH7, for some *hydrocarbon compounds to be degraded by microorganism*. Also, it contains a hydrophilic and a hydrophobic domain, may be facilitate the uptake of hydrocarbons into the producing cells

Application to Hydrocarbon Biodegradation

The high cell-retention capacity making CPVA and CPVAF polymers have an effective matrix, where they have the same condition, such as the polymer weight, time of incubation, pH, temperatures and the type of the cell. So, the high cell-retention capacities were expressed about the growth count of cell strains and not the potential of cell to degrade the oil. Where, this was attributed to adapting the cell to utilize some types of components present in oil.

Table 1: Percentage biodegradation of petroleum hydrocarbons after treatment with different bacterial strains in presence/ absence of each of biosurfactants and polymers (CPVAF & CPVA) separately

Sample	Weight of residual crude oil (g/l)	Percentage biodegradation* (%)
Control crude oil	1.0	-
B. lichenformis(B.l.)		

Crude oil + B.I.	0.15	85
Crude oil + B. I. + biosurfactant(1)	0.15	85
Crude oil + B. I. + CPVAF	0.29	71
Crude oil + B. I. + CPVA	0.12	88
Rhodococcuserythropolis(R.e.)		
Crude oil + R.e.	0.14	86
Crude oil + R.e+ biosurfactant(2)	0.12	88
Crude oil + R.e.+ CPVAF	0.30	70
Crude oil + R.e.+ CPVA	0.18	82
Pseudomonas xanthomarina (P.x.)		
Crude oil + P.x.	0.17	83
Crude oil + P.x.+ biosurfactant(3)	0.13	87
Crude oil + P.x.+ CPVAF	0.20	80
Crude oil +P.x.+ CPVA	0.10	90

*percentage biodegradation = Weight of original oil – wt. of residual / wt. of original oil * 100

As shown at Table 1, the percentage of biodegradation of petroleum hydrocarbon samples, were arranged in this order, R.e. >B.I. > P.x. and Bio2> Bio 3 >Bio1, with respect to free cells and their produced biosurfactants respectively. The capabilities of B.I. and P.x. to oil degradation were enhanced at immobilized them on CPVA and restricted with CPVAF. No further degradation of petroleum hydrocarbon was observed at immobilized R.e. cell on CPVA and its degradation ability was restricted with CPVAF. The data obtained from the degradation of petroleum hydrocarbon by immobilized cells in all the matrices suggest that the rate of degradation was much higher than that with freely suspended cells. This may be due to a kind of membrane stabilization, which is assumed to be responsible for the cell protection and better degradation rates in the adsorbed cells. However, the fact that the rate of degradation was less in the CPVAF foamed than in CPVA matrices, may be attributed to the slow diffusion of the compounds (oil)into the gel discs. And also, may be as a result of releasing of some cells from gel disc into the medium, because of the large size porosity of the gel matrices.

Growth Count of Microorganisms

Three different types of the bacterial strains were used to evaluate their potential for degradation of petroleum hydrocarbons. Their degradation efficiency was improved by each of, adding their producing biosurfactants, and also, by immobilized each of them on two different types of polymers, CPVA and CPVAF. The growth of the different bacterial strains inoculated into the mineral salt medium MSM on the crude oil as a sole carbon and energy sources in presence /absence each of the CPVA and CPVAF gels discs separately and also biosurfactants, were indicated in Table 2.

Table 2: Growth count of different bacterial strains, in the presence / absence of each of biosurfactants and polymers (CPVAF & CPVA) separately.

Sample	CFU/ml
B. lichemfomis (B.l.)	
Crude oil + B.l.	2.5×10^7
Crude oil + B. l. + biosurfactant(1)	3.5×10^7
Crude oil + B. l. + CPVAF	1.3×10^7
Crude oil + B. l. + CPVA	5×10^7
Rhodococcuserythropolis (R.e.)	
Crude oil + R.e.	5.6×10^7
Crude oil + R.e+ biosurfactant(2)	7×10^7
Crude oil + R.e.+ CPVAF	2×10^7
Crude oil + R.e.+ CPVA	5×10^7
Pseudomonas xanthomarina(P.x.)	
Crude oil + P.x.	4×10^7

Crude oil + P.x. + biosurfactant (3)	9×10^7
Crude oil +P.x. + CPVAF	7×10^7
Crude oil + P.x. + CPVA	8×10^7

The count was determined after incubated them at 30°C, 150 rpm, pH 7.5 for 15 days [28]. As shown in Table 2, the bacterial count on crude oil, was increased in these order where, R.e. >P.x.>B.l., but, when adding the biosurfactant (Bio), the order will be change to, P.x.> R.e.>B.l. Where the growth count of bacterial P.X. enhanced with adding its producing biosurfactant 3(Bio3), and also, when immobilizing it on each of CPVA and CPVAF. On other hand the growth count of bacterium B.l. and R.e. were inhibited or restricted by immobilizing the on the polymer CPVAF, but enhanced with CPVA in case bacteria B.l. and somewhat restricted with R.e.

The Degradation of Paraffins Compounds

The residual of the n-paraffin after biodegradation was identified using GC as shown in Table 3.

The UCM% was high (84.68) in control sample than of the most residual samples, suggesting that the control sample we obtained was exposed to the most chronic oil pollution. Also, was noticed that the most remaining percentages of the n-paraffin after biodegradation are higher than it's of the crude oil control, suggesting that occurring more biodegradation for UCM than that for paraffin compounds. The results revealed that the bacterium, Bacillus lichneformis ATCC 10716 (B.l.), *Rhodococcus erythropolis* ATCC 13260 (R.e.) *and Pseudomonas xanthomarina* KMM 1447 strains, that, were isolated from a biological waste-water, have the capability to utilizing paraffins as the sole source of carbon and energy and this capability was increased in this order, R.e.> P.x.>B.l. This is attributed to their efficiencies to degrade the non polar compounds (paraffin), but, the bacterial (B.l.) has the best degradation for unresolved complex mixture (UCM) with comparison with others bacterium. This ability increase when it was immobilized on each of CPVAF and CPVA polymers. On the other hand the paraffins biodegradations percentage, by R.e. and P.x. microorganisms were enhanced at adding to them their produced biosurfactants (Bio2 and Bio3) (19.05%, 19.77%, respectively).

Table 3: Distribution of carbon number in gas chromatogram of residual crude oil after treatment by different bacterial strains in presence/ absence of each of biosurfactants and polymers (CPVAF & CPVA) separately.

Sample	Percentage residual of paraffins after bio-treatment			Total area paraffins	Total area sample	(%)paraffins	(%) UCM
	C9-20	C21-30	C31-39				
Control crude oil	78.20	17.6	4.2	3477.34	2.27×10^4	15.32	84.68
B. lichenformis(B.l.)							
Crude oil + B.l.	18.04	74.47	7.49	9.54×10^4	4.14×10^5	23.04	76.96
Crude oil+B.l.+biosurf.(1)	22.74	68.19	9.07	6.24×10^4	2.94×10^5	21.22	78.78
Crude oil +B.l +CPVAF	30.61	68.72	0.67	6597.87	1.87×10^4	35.28	64.72
Crude oil+B.l + CPVA	81.67	16.37	1.96	1871.78	5829.88	32.11	67.89
Rhodococcuserythropolis(R.e.)							
Crude oil + R.e.	62.26	36.66	1.08	2388.68	2.013×10^4	11.87	88.13
Crude oil+R.e.+biosurf.(2)	31.35	51.06	17.59	7.79×10^4	4.09×10^5	19.05	80.95
Crude oil+R.e.+CPVAF	0	0	0	0	0	0	0
Crude oil + R.e.+CPVA	20.93	69.67	9.4	4×10^4	1.85×10^5	21.62	78.38

Pseudomonas xanthomarina(P.x.)							
Crude oil + P.x.	38.42	46.75	14.83	2.26×10^{4}	1.9×10^{5}	11.89	88.11
Crude oil+P.x.+biosurf.(3)	39.9	50.33	9.77	7.83×10^{4}	3.96×10^{5}	19.77	80.23
Crude oil+P.x.+ CPVAF	49	48	3	2458.47	1.3×10^{4}	18.91	81.9
Crude oil +P.x.+ CPVA	35.98	53.55	10.47	7.62×10^{4}	3.63×10^{5}	20.99	79.01

These percentages degradation increase at immobilized these cells on CPVA, but when R.e. was immobilized on CPVAF, the complete degradation for the paraffins and UCM, was determined to indicating that this polymer foaming type enhanced the biodegradation efficiency of R.e. to degrade completely the paraffin from C_9 to C_{39} and UCM. As we can see, the macro porous CPVAF carrier could provide sufficient space for microbial growth. Also, the polarity of the produced biosurfactants by B.l., R.e. and P.x. was increased in this direction, (Bio 1< Bio 2 <Bio 3).

The Degradation of PAHs

The brief identification and quantification of PAHs in the investigated of the crude oil remained after biodegradation and the corresponding control sample using HPLC, were shown in Table 4.

Table 4: HPLC analysis of Polynuclear aromatics hydrocarbons distribution in crude oil after treatment by different bacterial strains, in presence/absence of each of biosurfactant and polymers (CPVAF & CPVA) separately.

Sample	2-member rings of polyaromatics					3-member rings of polyaromatics			4-member rings of polyaromatics				
	Naph	Acena	Acenaph	Fluo	Total Conc. (%)	Ph.	Anth.	Total conc. (%)	Flu	Pyr	B (a) anth	Ch	Total conc. (%)
Control crude oil	0	0	0	0	0	0	0	0	67.77	0	0	2.67	70.44
B. licheniformis (**B.l.**)													
Control crude oil + B.l.	18.9	22.87	0	0	41.77	0	0	0	4.12	0	1.15	9.03	14.3
Control crude oil + B.l. + biosurfactant (1)	9.96	0	0	0	9.96	0	0	0	6.13	0	2.42	3.19	11.74
Control crude oil + B.l. + CPVAF	0	0	0	0	0	0	0	0	69.24	0	0	0	69.24
Control crude oil + B.l. + CPVA	0	87.00	0.81	0	87.81	1.39	0	1.39	0.10	0	0.66	0	0.76
Rhodococcus erythropolis (**R.e.**)													

Crude oil + R.e.	86.21	0	0	0	86.21	13.79	0	13.79	0	0	0	0
Crude oil + R.e. + biosurfactant (2)	7.83	0	2.13	0	5.7	20.28	20.28	0	0	0	0	0
Crude oil + R.e. + CPVAF	27.60	0	1.84	0	25.76	14.54	10.98	3.56	0	0	0	0
Crude oil + R.e. + CPVA	43.32	37.76	3.99	0.99	0.58	9.53	3.69	5.84	14.05	0	7.63	6.42
Psuedomonas xanthomarina(**P.x.**)												
Crude oil + P.x.	27.33	0.10	3.68	20.87	2.68	0.36	0.22	0.14	9.42	0	20.56	29.57
Crude oil + P.x. + biosurfactant (3)	1.09	0	1.09	0	0	2.15	0	2.15	85.83	0	85.83	0
Crude oil + P.x. + CPVAF	52.46	0	0	0	52.46	0	0	0	0	0	0	0
Crude oil + P.x. + CPVA	54.43	17.68	29.55	7.20	0	1.08	0	1.08	7.2	0	1.12	6.08

Sample	5-member rings of polyaromatics					6-member rings of polyaromatics		
	B (b) flu	B (k) flu	B (a) pyrene	Dib (a,h) anth	Total conc. (%)	B (g,h,i) perylene	Indeno(1,2,3,-cd) pyrene	Total conc. (%)
Control crude oil	0	18.77	10.79	0	29.56	0	0	0
B. licheniformis (B.l.)								
Control crude oil + B.l.	0	11.29	30.76	0.90	42.95	0	0.98	0.98
Control crude oil + B.l. + biosurfactant (1)	0	20.38	39.89	17.06	77.33	0.68	0.29	0.97
Control crude oil + B.l. + CPVAF	0	30.76	0	0	30.76	0	0	0
Rhodococcus erythropolis (R.e.)								
Crude oil + R.e.	0	0	0	0	0	0	0	0
Crude oil + R.e. + biosurfactant (2)	4.44	6.89	60.56	0	71.89	0	0	0
Crude oil + R.e. + CPVAF	6.36	2.82	0	48.68	57.86	0	0	0

Crude oil + R.e. + CPVA	0	0.1	31.86	0.66	32.62	0.33	0.15	0.48
Pseudomonas xanthomarina (P.x.)								
Crude oil + P.x.	2.6	3.23	4.83	0.99	11.65	0.60	0.49	1.09
Crude oil + P.x. + biosurfactant (3)	7.92	2.73	0.28	0	10.93	0	0	0
Crude oil + P.x. + CPVAF	17.57	29.97	0	0	47.54	0	0	0
Crude oil + P.x. + CPVA	0	0.69	36.60	0	37.29	0	0	0

The HPLC results showed that, the presence of 2,3 and 4,5,6- member rings of PAHs. In the control sample the percentage concentration of 4 -member rings is higher than that of the 5-member rings of PAHs. The results showed that the freely suspended *Rhodococcus erythropolis* (R.e.) can utilize 5-member rings of PAHs completely and degraded them to 3 and 4 -member rings of PAHs, Table 4. So, there no remaining of 5-member rings of PAHs was detected. On the other hand, the efficiency of *Pseudomonas xanthomarina* to degrade the 5-member rings of PAHs is less than that of the R.e. but, higher than that of B.I., where, the percentage concentration of the 5-member rings of PAHs is increase in presence B.I. than that in control sample. This result may be attributed to ability of this microorganism (B.I.) to degrade the 4 -member rings of PAHs that has certain polarity to lower 2-member rings of PAHs, that may be combined with each other's to form the 5-member rings of PAHs. So, the degradation arrangement of polar compound by microorganism increase in this direction, R.e. > P.x.> B.I. The efficient of these microorganisms were enhanced by immobilized them on different types of polymers (CPVA and CPVAF) or by adding their produced biosurfactants to them. Biosurfactants accumulated at interfaces, forming micelles that lowered the surface tension and thereby enhanced the solubility of poorly soluble compounds in water [36]. So, it was shown (Table 4), that the degradation efficiency of the B.I. to degrade the low polar compound (4-member rings PAHs), was enhanced by adding its produced biosurfactant (Bio1) to it. But its potential to degrade the 5-member rings of PAHs, is not recorded only; furthermore the percentage concentration of these compounds increased (77.33%) than with the B.I. alone. Also, this behavior of biodegradation was noticed with adding the produced biosurfactant (Bio2) to R.e. strain, resulting in restricted the utilizing of this bacteria for the 5 -member rings of PAHs. It was found that, the combination of P.x. with its biosurfactant 3 (Bio3), enhanced its capability to degrade the 5-member rings of PAHs to 2 -member rings of PAHs. Moreover, this combination made them to utilize on the 4-member rings of PAHs, where, they were superior in removing these compounds to residue of 1.09%. The enhancement or restricting of these biosurfactants towards the degradation capabilities of these strains to degrade the crude oil may be attributed to their structures. So, due to the structure of Bio3 contains many hydrophilic groups act as adsorbance sites for crude oil and also, can react with polar compounds than others (paraffin) resulting in degradation these compounds. At immobilizing the strain B.l. on each of, CPVA and CPVAF polymers, it was found that, with CPVAF no significant degrading of 4 and 5-member rings of PAHs, was observed. Although, the efficient of this strain enhanced when it was immobilized on CPVA to remove these PAHs and the remains was 0.76 and 10.04% of each 4 and 5-member rings of PAHs respectively. This immobilizing method was enhanced the potential of this strain to degrade the 4 and 5-member rings of PAHs to 2-member rings of PAHs. But, this enhancement biodegradation proper decrease at immobilizing the R.e. strain on CPVA to degrade the 4 and 5-member rings of PAHs to 2 and 3-member rings of PAHs with the concentration of 43.32 and 32.62% respectively. However, this potential of this strain to degrade the 4 -member rings to the 3 -member rings was increased with CPVAF polymer. It was noticed that, the residual of the 5 -member rings increase more than that in the control sample, where, it may be as a result of the combination of low polar compounds. Also, this biodegradation behavior was occurred at immobilizing the P.x. strain on these polymers.

This study revealed that the more efficient degradation of fluoranthen (Flu) (4 -member rings) was completely by using P.x. cell combined with its biosurfactant or immobilized it on CPVA polymer. The high removal was occurred by freely P.x. and immobilized each of B.l. and R.e. cells on CPVA polymer or combined each of them with their produced biosurfactants. Also, chrysene (Chr) was removed completely by freely R.e. cell or by immobilized each of B.l., R.e. and P.x. cells on CPVAF and also, B.l. on CPVA. The superior removal was recorded by freely P.x. The complete degradation of Benzo [a] Pyrene (BaP) (5-member rings) was observed by immobilized each of B.l., R.e. and P.x. cells on CPVAF polymer and by freely R.e. cell. But, the superior degradation, was occurred by combination of P.x with its produced biosurfactant (Bio3), where, it being amphipathic molecules of the biosurfactant (Bio3) with a hydrophilic and a hydrophobic domain, may be facilitate the uptake of hydrocarbons into cells. It was observed that, the complete removal of Benzo [K] fluranthene (5-member rings) was only, by freely R.e cell and the good degradation was occurred by immobilized each of B.l., R.e. and P.x. cells on CPVA polymer and also, by immobilized R.e. cell on CPVAF polymer. As well, it was revealed the good degradation of Benzo [K] fluranthene by combined each of R.e.and P.x. cells with their produced biosurfactants and by freely P.x cell. Surfactants can increase the surface area of hydrophobic materials, in water environment, thereby increasing their water solubility. Hence, the presence of surfactants may increase the efficiency of microbial to degrade the pollutants. It was revealed that the physicochemical stability of the PVA carrier was increased and the hydroxyl groups created a hydrophilic micro-environment for the metabolism of the immobilized microorganisms. Thus there is a potential for the development of microbial technology for the treatment of waste water containing aromatic hydrocarbons.

CONCLUSIONS

The immobilization methods make the cells stay alive for a long time in the immobilized stage when the conditions are optimal. Bioremediation using immobilized cells has been widely investigated for numerous toxic chemicals such as paraffins and aromatic hydrocarbons. In the present study, the CPVA and CPVAF macroporous carriers were prepared based on PVA with a pore-forming agent (calcium carbonate) with respect for CPVAF and each of was crosslinked with epichlorohydrin. Calcium carbonates as foaming agent has lower toxicity to yeast cells. The CPVA and CPVAF hydrogel discs were showed; the higher stability in water but CPVA has greater biological activity than CPVAF hydrogels that has more crosslinked structure. The CPVA and CPVAF carriers can serve as a suitable material for immobilized microorganisms, as they have the better chemical and thermal stability. For immobilization of bacteria, the CPVA carrier supports the higher biomass density and microbial activity yield than the CPVAF carrier. The maximum growth count was for P.x. strain immobilized on CPVA (8×10^7 CFU/ml) and also at, presence of the Bio 3 with P.x (9×10^7 CFU/ml). The higher growth count may attribute to the greater mass transfer of the substrate (and oxygen) between the bulk solution and the immobilized on CPVA carrier or combined with Bio 3 microorganisms. The rate of degradation of oil by immobilized cells in various matrices is compared with that by freely suspended cells, and combined with their produced biosurfactants with regards to oil degrading capacity at various initial oil concentrations. The longevity of the hydrocarbons degrading activity was by immobilized cells and by presence of biosurfactant with cells. These preparation methods of CPVA, CPVAF and biosurfactants were inexpensive, low in toxicity, and easily performed; they can serve as a promising and economical technique for immobilizing microorganisms. In addition, may be after being modified, the CPVA and CPVAF could serve as valuable carriers or supports in the biotechnology and biochemistry fields.

ACKNOWLEDGEMENTS

We are grateful for the grant obtained from the members in Central Analytical Lab, Department of Processes Development and Petrochemicals Department, Egyptian Petroleum Research Institute (EPRI).

REFERENCES

1.   RR Colwell, JD Walker, JJ Cooney (1977) Ecological aspects of microbial degradation of petroleum in the marine environment.CRC Criti Rev Microbiol5: 423-445.

2.   YM Moustafa (2004)Contamination by polycyclic aromatic hydrocarbons in some Egyptian Mediterranean coasts. Journal of Biosciences Biotechnology Research Asia 2: 15-24.

3.   EJ Baum (1978) in: HV Gelboin, POP Tsâ ™o (Eds.), Polycyclic Hydrocarbons and Cancer, Environment Chemistry and Metabolism, Vol. 1, Academic Press, New York, San Francisco, London.

4.   KSM Rahman, Rahman JT, Lakshmanaperumalsamy P, Banat IM (2002)Towards Efficient Crude Oil Degradation by a Mixed Bacterial Consortium.BioresourTechnol 85: 257-261.

5.   Shuttleworth KL, CernigliaCE (1995) Environmental aspects of PAH biodegradation.ApplBiochemBiotechnol 54: 291-302.

6.   Wattiau P, in: S.N. Agathos, W. Reineke(2002)(Eds.), Biotechnology for the Environment: Strategy and Fundamentals, Kluwer Academic Publishers., Netherlands.

7.   Larkin MJ, Kulakov LA, AllenCCR (2005) Environ Biotechnol, Poland.

8.   Georgiou G, Lin SC, Sharma MM (1992) Surface-active compounds from microorganisms. Biotechnology 10: 60-65.

9.   Morikawa M, Hirata Y, Imanaka T (2000) A Study on the structure-function relationship of lipopeptidesbiosurfactants. BiochimicaetBiophysicaActa1488: 211-218.

10. Benincasa M, Contiero J, Manresa MA, Moraes IO (2002) Rhamnolipid production by *Pseudomonas aeruginosa*LBI growing on soap stock as the sole carbon source. Journal of Food Engineering 54: 283-288.

11. IM Banat (1995)Biosurfactants production and possible uses in microbial enhanced oil recovery and oil pollution remediation: A Review BioTechn 51: 1-12.

12. Kim SH, Lim EJ, Lee SO, Lee JD, Lee TH (2000) Purification and characterization of biosurfactants from Nocardia sp. L-417. BiotechnolApplBiochem31: 249-253.

13. Rosiak JM, Janik I, Kadlubowski S, Kozicki M (2003) Nano-, micro- and macroscopic hydrogels synthesized by radiation technique. NuclInstrAnd Meth B 208: 325-330.

14. Abdeen Z (2005)Ph.D.Thesis, Preparations and Applications of Some Friendly Environmental CompoundsAin-Shams University, Cairo.

15. Abdeen Z(2011)Swelling and Re-swelling Characteristics of Cross-linked Poly (vinyl alcohol)/Chitosan Hydrogel Film. J Dis SciTechn 32: 1337-1344.

16. AL-SabaghAM, AbdeenZ (2010) Preparation and Characterization of Hydrogel based on Poly Vinyl Alcohol Crosslinked by Different Crosslinker Used to Dry Organic Solvents. JPolym Environ18: 576-583.

17. RatnerBD, BryantSJ (2004) Biomaterials: where we have been and where we are going. Annu Rev Biomed Eng 6: 41-75.

18. Figueiredo KCS, Alves TLM, BorgesCP (2009)Poly(vinyl alcohol) films crosslinked by glutaraldehyde under mild conditions. J ApplPolymSci 111: 3074-3080.

19. Bezbradica D, Obradovicb B, Leskosek-Cukalovic I, Bugarski B, Nedovic V (2007) Immobilization of yeast cells in PVA particles for beer fermentation. Process Biochem 42: 1348-1351.

20. Sanchez-Chaves M, Ruiz C, Cerrada ML, Fernandez-Garcia M (2008) Novel glycopolymerscontainingaminosaccharide pendant

groups by chemical modification of ethylene-vinyl alcohol copolymers. Polym49: 2801-2807.

21.   De Borba BM, Urbansky ET(2002) Performance of poly(vinyl alcohol) gel columns on the ion chromatographic determination of perchlorate in fertilizers. J Environ Monit 4: 149-155.

22.   Li YF, Bai X, Men XH, Yang LQ (2008)Macroreticular carrier based on poly(vinyl alcohol) foam and its application, China.

23.   Abd El-hady A, Abd El-Rehim HA (2004)Production of Prednisolone by Pseudomonas oleovorans Cells Incorporated Into PVP/PEO Radiation Crosslinked Hydrogels.J Biomed Biotechnol 4: 219-226.

24.   Karel SF, Libik B, Roberstson CR (1985)The immobilization of whole cells: Engineeringprinciples.ChemEngSci 40: 1321.

25.   Kuyukina MS, Ivshina IB, Yu Gavrin A, Podorozhko EA, Lozinsky V I, et al. (2005)Immobilization of hydrocarbon-oxidizing bacteria in poly(vinyl alcohol) cryogelshydrophobized using a biosurfactant.J Microbiol Meth 65: 596-603.

26.   Haghighat S, Akhavan A, Assadi MM, Pasdar SH (2008) Ability of indigenous *Bacillus licheniformis* and *Bacillus subtilis*in microbial EOR. Int J Environ Sci Tech 5: 385-390.

27.   Haddad NI, WangJi, Bozhong Mu (2008)Isolation and characterization of biosurfactant producing strain, Brevibacilibrevis HOB1 J IndMicrobiolBiotechnol 35: 1597-1604.

28.   El-Sheshtawy HS (2011)Biosynthesis and evaluation of some biosurfactants potentially active in remediation of petroleum pollution Ph.D. Thesis.Maicrobiol.Dept. Faculty of science.Cairo University

29.   Benson HJ (1995) Microbiological applications: A laboratory manual general microbiology, (6th Edition), Wmc. Brown Company Publshers, PubuqueLowa, USA.

30.   Institute of Petroleum (1995)Characterization of pollutants-High resolution gas chromatography method.IP standard methods for analysis and testing of petroleum and related products, The Energy Institute.

31.   .Ferrus R, Pages P (1977) Determination of the water retention value (WRV) of chitosan, Cell. Chem. Technol.

32. Aboustate MA, Moustafa YM (2005) Biodegradation of polycyclic Aromatic Hydrocarbons from Petroleum Crude Oils, Egyptian Journal of Biotechnology.

33. Lal BHM, Khanna S (1996) Degradation of crude oil by AcinetobactercalcoaceticusandAlcaligenesodorans.   J ApplBacteriol 81: 355-362.

34. Hatch AC, Burton GA (1999) Photo-induced Toxicity of PAHs Hyalella Azteca and ChironomusTentans: Effect of Mixures and Behavior, Environmental Pollution 106: 57-167.

35. .Biermann M, Lange F, Piorr R, Ploog U, Rutzen H, et al. (1987) In: Falbe, J. (Ed.), Surfactants in Consumer Products, Theory, Technology and Application.

36. Kuiper I, Ellen L, Lagendijk RP, Jeremy PD, Gerda EML, et al. (2004) Characterization of two Pseudomonas putidalipopeptidebiosurfactants, putisolvin I and II, which inhibit biofilm formation and break down existing biofilms. MoleculMicrob 51: 97-113.

Chapter
6

Environmental Radioactivity of Te-Norm Waste Produced From Petroleum Industry in Egypt: Review on Characterization and Treatment

M. F. Attallah[1], N. S. Awwad[1],[2] and H. F. Aly[1]

[1]Hot Laboratories and Waste Management Center, Atomic Energy Authority, Cairo,

[2]Chemistry Department, Faculty of science, King Khalid university, Abha, KSA

INTRODUCTION

At present time the different environmental compartments suffer from excessive accumulation of various toxic pollutants, hazardous fallout contaminants and several naturally occurring radionuclides, including potassium-40, thorium and uranium with the natural decay series of Th and U as well as several other man-made radionuclides. It is now of common practice to regulate any uncontrolled release of hazardous wastes in different environmental compartment [1]. Toxic hazardous wastes are defined as containing chemicals posing substantial hazards to human health or to the environment when improperly treated, stored, transported, or disposed. Scientific studies show that these wastes have toxic, carcinogenic, mutagenic, or teratogenic effects on human or other life forms. The majority of hazardous waste is generated by the chemical manufacturing, petroleum, pesticides and coal processing industries. Hazardous wastes may enter the body through ingestion, inhalation, absorption, or puncture wounds [2].

Naturally Occurring Radioactive Materials (NORM's) are those materials that contain radioactive elements what are found naturally in the earth's environment. Examples of these radioactive elements are the ^{238}U, ^{235}U, ^{232}Th series and their respective decay daughter, as well as ^{40}K. NORM's exist in soil, water, plants, animals, human, coal, lignite, petroleum, phosphate ores, geothermal wastes, wastewater... etc., in small but varying amounts almost everywhere [3].

On the other hand, nearly all the naturally occurring radioactive materials are considered in balance state. However, in several industrial processes e.g., mining of minerals (U, Th, steel, rare earth's metals), phosphate, oil and gas production, concentration of the natural radionuclides may be altered than its physical state, and exists in concentrations over than that exists naturally.

Wastes associated with the various industrial activities, with enhanced levels of the natural radioactivity as a result of industrial process, causes what is called, "Technological Enhanced-Naturally Occurring Radioactive Materials", to be named as acronym word "TE-NORM". For instance, TE-NORM scales may build up inside of oil field production tubing and may concentrate considerable quantities of radioactive material that has the potential to expose humans to relatively high dose of radioactivity. TE-NORM is often precipitated as sludge and scales. The human body cannot sense or detect TE-NORM, so, it can be detected and measured indirectly through their ionizing radiation using specialized instrumentation [4]. Uncontrolled release of activities associated with enhanced levels of NORM can contaminate the environment and pose a risk to human health. These risks can be alleviated by the adoption of controls to identify where NORM is present; and by the control of NORM-contaminated equipment and waste while protecting workers.

DISCOVERY TE-NORM IN INDUSTRY

The history of TE-NORM in oil and gas production follows closely in history of the discovery of radioactivity in the first part of 20th century. We must remember that, the discovery of radioactivity is more than one hundred years old. In 1918, a Canadian paper was published on radioactivity in natural gases. In the 1930's an elevated radium level were detected in the Russian oilfields. In 1953, the US geological society published a paper on uranium and helium in gas formations. In 1973, (EPA) performed a study on the presence Rn-222 natural gases [5-8].

A number of major oil companies helped sponsor a study on radon in natural gas products that was completed in 1975. The thrust of this study was the potential effect that radon would have on the consumer of natural gas products. Radon contamination of natural gas has been known for nearly 100 years [9]. These studies concluded that, radon in natural gas products does not present any hazard to the consumer. However, that radon could be a problem for different processing industries and some research efforts have focused on this concern [9-10].

In 1981, scale produced on offshore oil platforms in the North Sea was found to contain TE-NORM in significant quantities. These findings were presented in a 1985 offshore technology conference paper on radioactive scale formation in Houston, Texas, USA. Consequently, industry and government officials were aware of the possibility that TE-NORM scale could be present in US domestic operations. In 1986, significant TE-NORM scale was found in Laurel, Mississippi (USA). Some rather alarming press headlines and featured articles followed shortly in both Mississippi and Louisiana after the presence of TE-

NORM in the oil path became better known. These articles stressed the fact that there are no current regulations, either by USA or other federal governments controlling this radioactive waste and called for their creation and enforcement. Since TE-NORM was first re-discovered domestically, the oil and gas industry has responded progressively to TE-NORM issues by notifying appropriate state agencies, initiating field surveys and studies to characterize and locate occurrence of TE-NORM in conjunction with the American Petroleum Institute, informing other oil and gas operations, employees and contractors, and reviewing operating practices [5].

TE-NORM contamination of oil and gas industry petroleum equipment has been identified world- wide, e.g. USA (Alaska, Gulf of Mexico region), the North Sea region, Canada, Australia, several Middle East countries (Egypt, Saudi Arabia...etc.). Since 1918 till 1980, most researches were focused on the TE-NORM contamination of natural gas facilities, and the contamination is attributed to ^{226}Ra as well as ^{222}Rn gas and its decay products, e.g. ^{218}Po, ^{214}Pb, ^{214}Bi, and ^{210}Pb. Within the text, the activity concentration ranges from background level to several hundreds Bq/g (^{226}Ra). Doses to workers involved in handling, contaminated equipment, or waste are usually very low, and the main problem related to radioactive deposits is waste disposal [11]. The presence of TE-NORM or naturally occurring radionuclides in the product materials from oil and gas facilities, give rise to deposits with enhanced levels of these radionuclides in the processing equipment [12].

ORIGIN AND FORMATION OF TE-NORM

In nature, there are three naturally occurring radioactive decay series. The first series, known as thorium series, consists of a group of radionuclides related through decay in which all the mass numbers are evenly divisible by the number four, (4n) series. This series has its origin radionuclides Th-232, its abundance is 100%, specific activity is 2.4×10^5 dpm/g, which undergoes α -decay with a half-life of 1.41×10^{10} y. The terminal nuclide in this decay series is the stable Pb-208. In this series, the transformation from the original parent Th-232 to the finial product Pb-208 requires 7α and 4β -decays. The long-lived intermediate is 6.7 y for Ra-228, Fig. (1).

The second series is the uranium series, which consists of group radionuclides that, the parent radionuclide in this series is U-238 $(4n^{+2})$ abundance = 99.27 %, which undergoes α -decay with a half-life of 4.47×10^9 y. The stable product of the uranium series is Pb-206, which is reached after 8α and 6β -decay steps, Fig. (2). This is a particularly important series in nature since it provides the more important isotopes of elements Ra, Rn and Po, which can be isolated in large amounts in the processing of uranium minerals. Each ton of uranium is associated with 0.34 g of Ra-226.

The third radioactive decay series, $(4n^{+3})$ known as the actinium series, the head of this series is U-235, which has an abundance of 0.72 % and a half-life of 7.1×10^8 y for α -decay. The stable end product of this series is Pb-207, which is formed after 7α and 4β -decay steps. The specific activity of U-235 is 4.8×10^6 dpm/g, Fig. (3) [13].

Other important radionuclide that exists in the nature is potasium-40 (t1/2. 1.28×10^9 y, isotopic abundance 0.0118 %). K-40 is found in plants, animals and in human bones. It is widely distributed in nature with volume concentrations ranging from 0.1 to 3.5 % in carbonates (limestones). The bones of an average human body contain concentrations of ~ 17 mg of K-40. The average radiation dose received from K-40 is 0.25 mSv/y to tissue and ~ 0.36 mSv/y to bone.

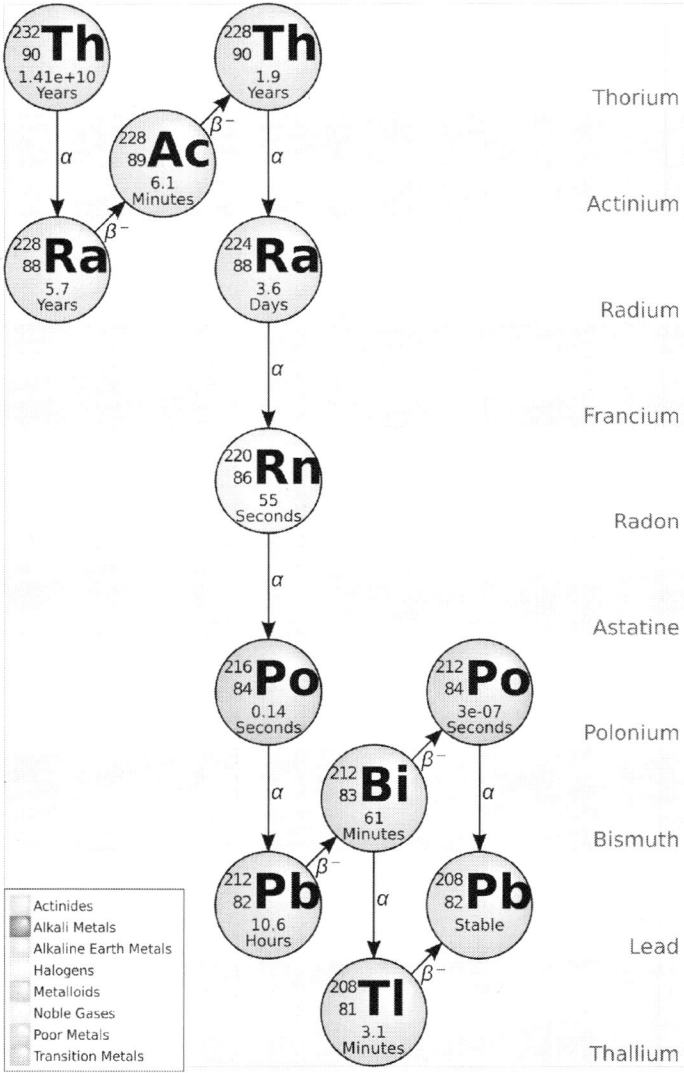

Figure 1: Scheme of the thorium decay (Th-232) series.

although the activity concentration of [226]Ra in the MS is higher. The variation of EF is independent of the [226]Ra content and is strongly correlated to the grain surface density [7,8,19]. The smaller the grain size the higher the EF [26].

The main types of scale encountered in oil & gas facilities are sulphate scale which results from drilling clay, it called Barite slurry. It is usually colorless or milky white, but can be almost any color, depending on the impurities trapped in the crystals during their formation. Barite is relatively soft, measuring 3-3.5 on Mohs' scale of hardness. It is unusually heavy for a non-metallic mineral. The high density is responsible for its value in many applications. Barite slurry is generally used as mud in drilling oil. Barite is chemically inert and insoluble. Radium is chemically similar to barium (Ba), strontium (Sr) and calcium (Ca), hence radium co-precipitates with Sr, Ba or Ca scale forming radium sulphate, radium carbonate and – in some cases – radium silicate. The mixing of seawater, which is rich in sulphate, with the formation water, which is rich in brine, increases the scaling formation tendency. Also the sudden change in pressure and temperature or even acidity of the formation water, as it is brought to the surface, contributes to scale build-up. This phenomenon has significant implications for the production of oil; in this case the capacity of the pipe to transfer oil would be reduced significantly. The activity concentration of [226]Ra and [228]Ra in hard scales in Egypt and some other countries are mentioned in Table 2.

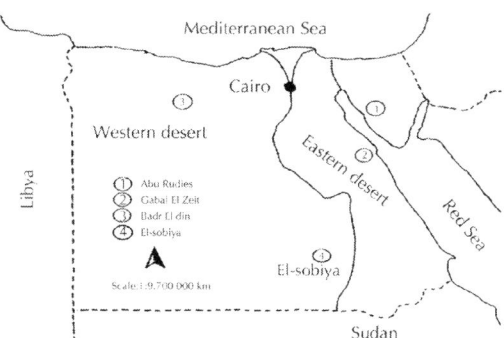

Figure 4: Map for some sites of the TE-NORM wastes associated with phosphate and petroleum and gas production in Egypt.

El Afifi and Awwad [27] characterized TE-NORM from Abu Rudeis region (onshore oil field, Suez gulf area) in the North Sinai Governorate, Egypt. The mineralogical analysis by X-ray techniques (XRF and XRD) has been carried out. Table 3 represents the chemical analysis of the TE-NORM waste samples using the XRF technique. The data showed major elements (Si, Fe, Al and Na) and alkaline earth elements (Mg, Ca, Sr and Ba) [27].

Table 2: Activity of ^{226}Ra (U-series), and ^{228}Ra (U-series) in the TE-NORM in Egypt and some other countries [6, 21, 27-33].

Country/region	Activity (Bq/g)		Ref.
	226R$_a$	^{228}Ra	
Egypt/oil field			
Abu Rudeis	68.9	24	[27]
Gabal El Zeit	14.8	4.3	[28]
Badr El Din	31.4	43.3	[28]
Red Sea	195	897.8	[29]
Western desert	59.2	244.5	
South Sinia	6.99	1-1.9	
South Sinia	506	32-50	[30]
Other countries			
Australia	20-70		
USA	70.8		[21]
Algeria	1-950		
Tunisia	4.3-658		
UK	1-1000		[6]

The radioactivity of naturally occurring radionuclides of ^{238}U, ^{235}U, ^{232}Th-series and ^{40}K in the sediment samples of the TE-NORM waste from Abu Rudeis region before fractionation (B.F.) and after fractionation (A.F.) are given in Table 4. The bulk waste was fractionated into nine homogeneous fractions with different particle sizes (< 3.0– < 0.1 mm) to show the effect of particle size on the activity distribution. Moreover, radiation hazardous indices including the radium equivalent activity (Ra-eq), radon (^{222}Rn) emanation coefficient (EC) and absorbed dose rate (Dγr) were also estimated of TE-NORM waste. The radon emanation coefficient (EC) is a very important radiological index used to evaluate the amount of the 222Rn emanated fraction released from the waste materials containing naturally occurring radionuclides (e.g. ^{238}U, ^{235}U and ^{226}Ra). In this study, the assessment of Rn EC is related only to ^{222}Rn decayed from its parent ^{226}Ra content in the waste. Since ^{222}Rn and its respective decay progenies (e.g. ^{210}Pb and ^{210}Po) have longer half-lives than that of other radon isotopes, ^{222}Rn is considered to be more radiological hazardous to human health than radionuclides coming from other radon isotopes. Fig. 5 represents the ^{222}Rn EC released from the bulk TE-NORM waste and the different fractions. The amount of the ^{222}Rn fraction emanated from the bulk waste was 0.066. It is found that the grain size has an effect on the amount of ^{222}Rn EC. There is a gradual increase in the ^{222}Rn EC with the waste particle size. This was observed in fine particle sizes from less than 0.1 mm up to 2 mm. In this range, ^{222}Rn EC increased from 0.041 to 0.086. There is no effect of the waste particle sizes on the ^{222}Rn EC released from large particle sizes as shown in the grain sizes between 2 and 3 mm. The radon EC found in this range of particle sizes was ranging between 0.093 and 0.095 [27].

Table 3: Results of XRF analysis of the TE-NORM waste before fractionation and after sieve fractionation for some selected fractions [27].

Sample code	Concentration (%)							
	Mg	Ca	Sr	Ba	Na	Al	Fe	Si
B.F.a	2.44	13.53	0.91	3.25	4.13	3.63	27.12	44.05
A.F.b								
F2 (0.1-0.2) mm	2.51	17.94	0.69	2.14	3.82	4.66	29.15	48.75
F4 (0.3-0.5) mm	1.93	11.23	1.27	4.63	4.22	3.67	29.69	44.33
F6 (1.0-1.6) mm	1.53	7.78	1.53	5.47	4.00	3.76	29.96	47.21
F8 (2.0-2.6) mm	1.31	8.65	1.66	6.37	4.58	3.91	27.61	47.39

Reported levels of the ^{226}Ra and ^{228}Ra activity concentrations observed in the solid scale and sludge in different countries are listed in Table 5.

Table 4: Results of gamma-spectrometric analysis of the TE-NORM waste before (B.F.) and after (A. F.) dry fractionation [27]

| Sample | ^{238}U-series | | ^{235}U-series | | ^{232}Th-series | | | 4 0 K |
	2 3 8 U (Bq/g)	2 2 6 R a (Bq/g)	2 1 o p b (Bq/g)	2 2 3 R a (Bq/g)	2 2 8 A C (Bq/g)	2 1 2 P b (Bq/g)	2 0 8 T l (Bq/g)	(Bq/g)
B.F.	7.1	86.9	4.4	2.7	24.0	22.4	25.2	1.3
A.F.*								
F1	7.5	60.4	4.3	3.5	25.8	24.7	22.9	1.9
F2	4.5	43.0	4.2	1.7	19.1	17.0	20.3	1.1
F3	6.2	55.4	4.3	2.6	21.5	21.0	20.5	1.3
F4	9.2	81.1	5.3	3.7	34.6	33.3	36.5	2.5
F5	11.2	96.5	4.3	5.2	41.4	37.2	39.0	3.4
F6	9.6	85.4	4.4	3.5	39.7	36.2	40.1	3.6
F7	7.1	78.3	4.5	4.0	35.4	36.4	32.9	4.3
F8	11.8	102	6.5	5.7	43.7	38.5	39.0	4.8
F9	7.1	54.5	2.7	3.4	22.8	20.1	22.1	2.5

The radium equivalent activity (Ra-eq) is a radiation index, used to evaluate the actual radioactivity in the materials containing naturally occurring radionuclides, e.g. ^{238}U and ^{232}Th series, and/or 40K. Values of Ra-eq activity for the bulk TENORM waste and the waste fractions were calculated. It is clear that Ra-eq exceeds the maximum permissible radium activity (Ref. value is 370 Bq/kg) as reported by Zaidi et al. [23]. The higher Ra-eq activity reached about 164.9 kBq/kg in fraction F8, whilst the lower value amounted 70.4 kBq/kg in F2 (Fig. 6). The higher the radioactivity level in the waste, the higher is the radiological impacts, especially when considering the potential of operators to be exposed via internal contamination by ingesting the dust during waste processing. The total absorbed dose rate due to g-emissions was estimated and the obtained values are presented in Fig. 6 [27]. It is recommended that the acceptable total absorbed dose rate by the workers in areas containing g-radiations from ^{238}U and ^{232}Th series and their respective decay progenies, as well as 40K, must not exceed 0.055 mGy/h [24]. It is obvious that the calculated total absorbed dose rates for all waste samples are higher than the recommended dose level that are acceptable (Fig. 8). The low total absorbed g-dose rate is 31 mGy/h in fraction F2, while the high value is 72.7 mGy/h in fraction F8. It is clear that the absorbed dose rate depends on the activities of g-emitters (e.g. ^{226}Ra, ^{232}Th ^{40}K), while it is independent of the waste particle size. Therefore, the total absorbed dose rates increases with the activity concentration, and consequently enhances the radiological impact on the workers surrounded by the wastes.

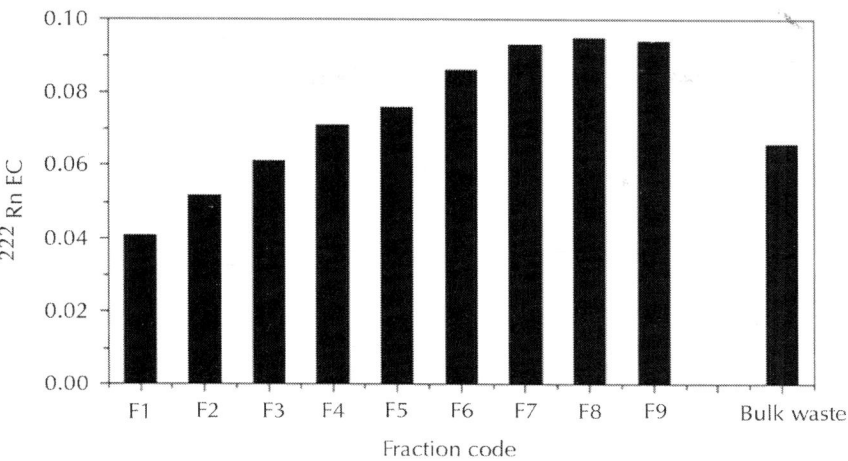

Figure 5: Effect of the waste particle size on the radon emanation fraction release on TENORM of petroleum in Egypt.

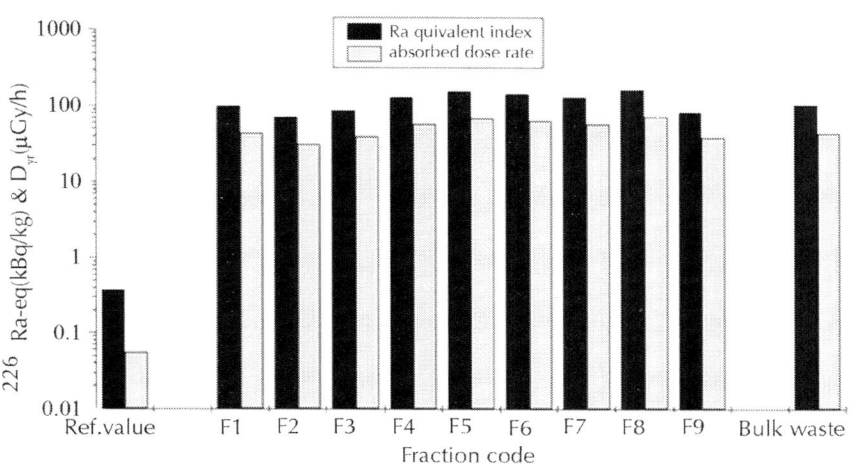

Figure 6: Variation of radium equivalent index and total absorbed dose rate with waste fractions and bulk Waste of TENORM of petroleum in Egypt.

Table 5: Ranges of activity levels of 226[Ra] in different scale and sludge samples

Field	Sample	²²⁶Ra (Bq/kg)	²²⁸Ra (Bq/kg)
Algeria [32]		1000-950.000	
Australia [34]		21.000-250.000	48.000-300.000
Brazil 135]		19.100-323.000	4210-235.000
Brazil [36]		121.000-3.500.000	148.000-2.195.00
Brazil [37]		77.900-2.110.000	101.500-1.550.00
Congo [38]		97-151	
Egypt (271		68.900	24.000
Egypt [29]		7541-143.262	35.460-368.654
Egypt (30]		493-519	32-50
Italy [38]		< 2.7-2890	
Kazakhstan (39]		510-51.000	200-10.000
Malaysia [40]		114.300-187.750	130.120-206.630
Norway [41]		300-32.300	300-33.500
Saudi Arabia [42]		.08-1.5	
Tunisia [38]		31-1189	
Tunisia [33]		4300-658.000	

USA [43]		1000-1.000.000	
UK [6]		3.700.000	
USA [21]		15.400-76.100	
Australia [34]		25.000	30.000
Brazil [35]		50.000-168.000	49.000-52.000
Brazil [36]		< LLD-413.000	< LLD-117.900
Egypt [29]		18.000	13.250
Egypt [30]		5.27-8.68	1-1.9
Malaysia [40]		6-560	4.520
Norway [41]		100-4700	100-4600

As shown in Table 5, the concentration levels of radium nuclides in scale vary within a wide range being much higher than those of the sludge. According to the latest Environmental Protection Agency (EPA) estimation, the average radium nuclide concentration is around 18,000 Bq/kg and 2800 Bq/kg in scale and sludge, respectively [44]. Elevated concentration activities of both radionuclides, exceeding the exemption level of 10,000 Bq/kg recommended by IAEA safety standards, were frequently found in the scale samples. A large uncertainty is observed in the estimations of the total amount of radioactive waste generated by oil industry, and the EPA assumes that 100 tons of scale per oil well are generated annually in the United States [44], while for the North Sea wells a somewhat lower value of 20 t is suggested [45] and only 2.25 t per year by one oil-producing well for Latin American oil producing countries [46]. It was also estimated that approximately 2.5×10^4 and 2.25×10^5 tons of contaminated scale and sludge, respectively, were generated each year from the petroleum industry in the middle of the previous decade [47,48]. This means that TENORM waste from the oil industry may generate radiation exposure levels which require attention and continuous monitoring during some routine operation in this industry. This exposure is caused by external radiation coming from the ^{226}Ra radionuclide and its progenies: ^{214}Pb and ^{214}Bi as well as by inhalation of α -emitting radionuclides:^{222}Rn as well as ^{218}Po and ^{214}Po formed from ^{222}Rn escaping into the air adjacent to scale deposits, as reported in Table 6.

Table 6: Exposure rate levels in the oil industry

Country	Reported range (pSv/h)
Algeria [32]	Back Ground-100
United Kingdom [32]	10-300
Egypt [27]	50-100
Congo, Italy, Tunisia [38]	0.1-6
USA [49]	Back Ground-100

Te-Norm in Produced Water

Oil and gas reservoirs contain water (formation water) that becomes produced water when brought to the surface during hydrocarbon production. Oil reservoirs can contain large volumes of this water whereas gas reservoirs typically produce smaller quantities. In many fields, water is injected into the reservoir to maintain pressure and / or maximize production. In many offshore oil fields sea water is additionally injected to maintain pressure, and it mixes with the formation water. In such cases, in the exploited oil/water mixture, the content of the production water can reach even 95%. For this reason, the produced waters are typically saline and rich in Cl$^-$ anions forming aqueous complexes with Ra that enhance the mobility of Ra nuclides from adjacent geological rocks into these waters [50]. Comprehensive older literature reviews of radium nuclide concentrations in formation and produced water indicated an average radium nuclide concentration in waters in excess of 1.85 Bq/dm^3 and exceptionally up to ~ 1000 Bq/dm^3[49,51,52]. As ^{226}Ra originated from the radioactive decay of

^{238}U, while ^{228}Ra from ^{232}Th, the ^{226}Ra/^{228}Ra ratio in the oil-field brines depends on the U/Th ratio of the reservoir rock and ranges from 0.1 to 2.0, but for the most cases its activities are comparable.

Typical ranges or average values of the radium radionuclide concentrations in the formation or produced water from different oil fields, including the recent data, are listed in Table 7 [29,32,34,38,51,53-62].

A critical review of the intense studies of the activity concentrations of ^{226}Ra, ^{228}Ra as well as ^{210}Pb and ^{210}Po in produced water in 2003 from Norwegian oil and gas platforms located in the North Sea were also reported [55]. The concentrations of ^{226}Ra and ^{228}Ra in produced water discharged from these offshore platforms vary between 0.1 Bq/dm^3 and about 200 Bq/dm^3 with the average values estimated to be 3.3 Bq/dm^3 and 2.8 Bq/dm^3, respectively. Slightly higher radium activities ~ 10 Bq/dm^3 have been found for produced water outfalls in the Gulf of Mexico [63]. The European Commission (EC) derived the specific clearance levels at low activity for metals and buildings in radiation protection. The worldwide average concentration of these radionuclides in produced water discharged to the environment is estimated at 10 Bq/l. These concentrations are approximately three orders of magnitude higher than natural concentrations of radium in drinking or sea water. Because the radium radionuclide concentrations in that waste water are usually below the clearance levels (Table 8), it is recognized as a low specific activity waste and they may be injected into underground formations or disposed into the sea.

Table 7: Ranges of activity levels in produced water from the oil fields

Field	Sample	^{226}Ra (13q/c1m1	^{228}Ra (Bq/dm^3)
Algeria [32)	Formation water	5.1-14.8	23'
Australia [34]	Produced water	17'	0.05-12
	Produced water	0.01-6	1-59
Brazil [53]	Produced water	5.1'	2.8'
Congo [38]	Formation water	5-40	0.5-21
Egypt [29)	Produced water	0.2-2	8.8-60.4
Italy [38)	Produced water	0.3-10.4	0.7-1.7
Norway [54]	Produced water	3.3'	15.P
Norway [55]	Produced water	0.5-16	25-30
Norway [56]	Produced water	9.9-111.2	
Syria [57]	Produced water	1.7'	
UK [58]	Produced water	0.1-60	
USA [51)	Produced water	0.15-21.6	
USA [591	Produced water	12.6'	
USA [60]	Produced water	22-30	
USA [61)	Crude oil	26.5-217.5	
Egypt [62)			

Table 8: The European Commission Clearance levels

Radionuclide	Concentration (Bq/g)
40K	100
226Ra	10
232Th	1

A comprehensive evaluation of discharges from the oil industry to the sea was done for European waters during the European Commission Marina Project [64].

The annual release of ^{226}Ra and ^{228}Ra with produced water from off-shore fields in Europe in the 1990s stabilized at around 5 and 2.5 TBq per year, respectively. The commonly used two steps model of the radionuclide dispersing and diluting in the water in the vicinity of the oil platforms predicts a diluting factor up to 10^3 within minutes and within a few meters of the discharge source [65]. Therefore, additional radium nuclide concentrations in seawater of the local zone could be estimated as equal to around 5–10 Bq/m^3, in comparison with the natural concentration of around 1 Bq/m^3 for ^{226}Ra. The radium activities in the produced water for the North Sea, a yearly release of 0.65 TBq for ^{226}Ra and 0.33 TBq for ^{228}Ra was appraised for offshore oil production from Argentina and Brazil [66].

Produced water contains a complex mixture of inorganic (dissolved salts, trace metals, suspended particles) and organic (dispersed and dissolved hydrocarbons, organic acids) compounds, and in many cases, residual chemical additives (e.g. scale and corrosion inhibitors) that are added into the hydrocarbon production process. Feasible alternatives for the management and disposal of produced water should be evaluated and integrated into production design. These alternatives may include injection along with seawater for reservoir pressure maintenance, injection into a suitable offshore disposal well, or export to shore with produced hydrocarbons for treatment and disposal. If none of these alternatives are technically or financially feasible, produced water should be treated before disposal into the marine environment. Treatment technologies to consider include combinations of gravity and / or mechanical separation and chemical treatment, and may include a multistage system, typically including a skim tank or a parallel plate separator, followed by a gas flotation cell or hydrocyclone. There are also a number of treatment package technologies available that should be considered depending on the application and particular field conditions. Sufficient treatment system backup capability should be in place to ensure continual operation and for use in the event of failure of an alternative disposal method, for example, produced water injection system failure. Where disposal to sea is necessary, all means to reduce the volume of produced water should be considered, including:

- Adequate well management during well completion activities to minimize water production;
- Recompletion of high water producing wells to minimize water production;
- Use of down hole fluid separation techniques, where possible, and water shutoff techniques, when technically and economically feasible;
- Shutting in high water producing wells. To minimize environmental hazards related to residual chemical additives in the produced water stream, where surface disposal methods are used, production chemicals should be selected carefully by taking into account their volume, toxicity, bioavailability, and bioaccumulation potential [67].

The average worldwide activity levels of uranium (U), thorium (Th) and potassium (K) [68] and the exemption activity levels of NORM as recommended in the IAEA basic safety standards [69], were given at Tables 9&10. The average worldwide levels of the most common radiological indices [68] was given at Table 11. These indices include radium equivalent (Ra-eq), total absorbed dose (D_{yr}) and effective annual dose rate (EDAR).

Table 9: The average worldwide activity levels of U, Th and K [68]

Radionuclide	U	Th	K
Activity level (13q/Kg)	50	50	500

Table 10: The exemption activity levels of NORM as recommended in the IAEA basic safety standards [69].

Radionuclide	U	Ra	Rn	Th	Ra	Ra
Exemption level (8q/g)	1	10	10	1	10	10

Table 11: The average worldwide levels of the most common radiological indices [68]

Radiological indices	Ra-eq **B**q/Kg	Dw (Unit)	EADR (mSv/yr) for worker	EADR (rtiSytyr) for Public
Activity level	370	55	20	1

INVESTIGATION OF TE-NORM TREATMENT IN PETROLEUM INDUSTRY

In the last decade, attention was focused on the environmental and health impacts from the uncontrolled release of TENORM wastes [19,28,70,71]. Therefore, the treatment of these wastes is of increasing interest because accumulation of large amounts with a significant activity may cause a health risks to the workers through exposure, inhalation of radon (^{222}Rn) decayed from radium and/or ingestion of waste dust during the periodical maintenance of the equipment used. The trials towards the treatment of TENORM wastes from many industries are still limited. In this concern, removal of ^{226}Ra from TENORM wastes produced from oil and gas industry was carried out by a simple extraction process using saline solutions (i.e., seawater) and chemical solutions [70]. The chemical treatment process of TE-NORM sludge has been carried out by suspending the clay fraction content in the solid waste in suitable leaching solutions was reported. It was found that, the maximum removal % of ^{226}Ra is ~ 85% [70]

El-Afifi, studied the treatment of radioactive waste containing ^{226}Ra from oil and gas production, using different chemical solutions, in terms of a simple and sequential techniques based on suspending ^{226}Ra through the clay fraction in the waste. More than 50% of ^{226}Ra was removed through the treatment using moderate acids and salts solutions, while more than, 75% of ^{226}Ra was removed based on successive treatment or using some strong chelating reagent solutions [28].

The development of the treatment of a sludge TENORM waste produced from the petroleum industry in Egypt, using selective leaching solutions based on two approaches 'A' and 'B' has been investigated by El-Afifi et al. [71]. The results obtained showed that treatment of the waste through main four successive leaching steps removed ~ 78 and 91% of ^{226}Ra, 65 and 87% of ^{228}Ra as well as 76 and ~ 90% of ^{224}Ra using approaches A and B, respectively [71]. El-Afifi et al. [72] reported some data about the radiological characterization for phosphogypsum waste and phosphate rocke samples by γ -ray spectrometer. El-Didamony et al.[73] used the solvent extraction technique in treatment of phosphogypsum waste obtained as a byproduct of phosphoric acid production from phosphate ore.

CONCLUSION

The naturally occurring radioactive materials (NORM) are found everywhere. We are exposed to it every day. NORM represent an integral part of the planet, our bodies, the food we eat, air we breath, the places where we live and work, and within products we use. However, in the exploration and extraction processes of oil and gas, the natural radionuclides ^{238}U, ^{235}U and^{232}Th, as well as the radium-radionuclides (^{223}Ra, ^{224}Ra, ^{226}Ra and ^{228}Ra) and ^{210}Pb,.etc., are brought to the slurry surfaces and may contain levels of radioactivity above the surface background. The petroleum waste (scale or sludge) have been produced by two mechanisms: either incorporation or precipitation onto the production equipment such as: pipelines, tank storage, pumps,..etc. The waste generated in oil and gas equipment is due to the precipitation of alkaline earth metals as sulfate, carbonates and/or silicates. It is clear that PG and sludge and scale wastes represent one of the major sources of ^{226}Ra in the environment. Since the concentration of ^{226}Ra found in both of them waste exceeds that permitted by the international regulations, it was found necessary to reduce the risks due to indoor radon and direct γ -radiation in each wastes to be used in different life aspects.

REFERENCES

1. A. Abed-Rassoul, Proc. Lectures, of 2nd Arab Conf. on the peaceful uses of atomic energy, Cairo, Egypt, 1 (1994

2. H.F.David Liu and G. Bela Liptak,"Hazardous waste and solid waste", Lewis Publishers, USA (2000

3. F. A. Shehata, E. M. El -Afifi, H. F. Aly, of. Proc, Conf. Int, in Radioactive waste management and environment remediation, Nagoya, Japan, CD-ROM, Sept. 26-30 (1999

4. M. F. Maged, E. A. Saad, Management. Environ, 9(2. Health, 1998

5. K. J. Grice, Proc, of the 1st Int. Conf. on Health, Safety and Environment, Hague, Netherlands, Nov. 10-14, 559 (1991

6. Exploration & Production Forum,1987Low specific activity scale origin treatment and disposal. Report 6Old Burlington Street, London WlX 1LB, 2528

7. R. Colle, R. J. Rubin, L. I. Knab, J. M. R. Hutchinson, N. B. S. Technical, No. note, 113, D. C. U. S. Washington, of. Department, National. Commerce, of. Bureau, Standards, 19811

8. B. TANNER, Radon migration in the ground. A supplementary review, in: Natural Radiation Environment III, Vol. 1, T. F. GESELL, W. M. LOWDER (Eds), US Department of Energy Report CONF-780422, 1980, p. 5.

9. F. S. Grimaldi, M. H. Fletcher, L. B. Jenkins, J. Anal, Chem, 2. , 1957

10. K. Sudhalatha, 1. Talanta, 1963

11. T. Strand, I. Lysebo, Proc, of the 2nd Int. Symp. on the Treatment of NORM, Krefeld, Germany, Nov. 10-13, 137 (1998

12. J. E. Oddo, X. Zhou, D. G. Linz, S. He, M. B. Tomoson, Proc, of Exploration and production environmental Conf., Houston, Texas, USA, March 27-29, 207 (1995

13. G. R. Choppin, J. , Rydberg,"Nuclear chemistry theory and application",st Ed., Pergamon Press, New York, USA, 65 (1980

14. Kadi. H. Hamlat, S. Djeffal, H. Brahimi, 2003Radon concentrations in Algerian oil and gas industryAppl. Radiat. Isot. 58125130

15.   P. J. Shuller, D. A. Baudoin, D. J. Weintritt, Proc, of Exploration and production environmental Conf., Texas, USA, March 27-29, 219 (1995

16.   Testa, D. Desideri, F. Guerra, M.A. Meli and C. Roselli, Proc. of the Int. Solv. Extraction Symp., Moscow, Russia, June 21-27, 416 (1998).

17.   K. P. Simth, D. L. Blunt, G. P. William, L. L. Tebes, Proc, of Exploration and production environmental Conf., Houston, Texas, USA, March 27-29, 231 (1995

18.   M.A.Hilal,"Nuclear spectroscopic measurements for materials and environmental control", M. Sc. Thesis (Physics), Faculty of Science, Zagazig University (Banha branch), Egypt (1998

19.   M. F. Attallah, Chemical studies on some radionuclides in industrial wastes. M. Sc. Thesis, Benha University, Egypt (2006).

20.   S. Stoulos, M. Manolopoulou, C. , Papastefanou,"Measurement of radon emanation factor from granular samples: effects of additives in cementJ. App. Radiat. Isot., 60, 49 EOF54 EOF2004

21.   G. J. White, A. S. Rood, emanation. Radon, N. O. R. M. from, pipe. contaminated, scale, at. soil, industry. petroleum, sites, J. White, A. S. Rood, Radon emanation from NORM contaminated pipe scale and soil at petroleum industry sites. J. Environ. Radioact., 5420012001401413

22.   J. Beretka, P. J. Mathew, J. Health, Society. 4. Phys, 1985

23.   J. H. Zaidi, M. Arif, S. Ahmed, S. Fatima, I. A. Quresh, 1991Determination of natural radioactivity in building materials used in the Rawalpindi/Islamabad area by g-ray spectrometry and instrumental neutron activation analysis. Applied Radiation and Isotopes 51559564

24.   UNSCEAR, United Nations Scientific Committee on the Effect of Atomic Radiation: Sources and effects of ionizing radiation, United Nations, New York (1993).

25.   UNSCEAR, United Nations Scientific Committee on the Effect of Atomic Radiation: Sources and effects of ionizing radiation, United Nations, New York (2000).

26.   E. M. El Afifi, S. M. Khalifa, H. F. Aly, of. Assessment, 2. the, content. Ra, 2. the, fraction. emanation, T. E. of-N, O. R. M. wastes, certain. at, of. sites, petroleum, production. gas, Egypt. in,

of. Journal, Radioanalytical, Chemistry. Nuclear, Vol, 12004221 EOF224 EOF

27. E. M. El Afifi, N. S. Awwad, 2005Characterization of the TENORM waste associated with oil and natural gas production in Abu Rudies, Egypt, J. Environ. Radioact. 82719

28. E. M. El Afifi, 2001Radiochemical studies related to environmental radioactivities. Ph.D. thesis(Chemistry), Faculty of Science, Ain Shams University, Cairo, Egypt, 98.

29. S. Shawky, H. Amer, A. A. Nada, Maksoud. T. M. Abdel, N. M. Ibrahim, 2001Characteristics of NORM in the oil industry from eastern and western deserts of EgyptApplied Radiation and Isotopes 55135139

30. M. Abo-Elmagd, H. A. Soliman, A. Kh, N. M. Salman, Radiological. El -Masry, of. T. E. hazards-N, O. R. M. in, wasted. the, pipes. J. petroleum, Envron. Radio. 101, 51-54 (2010

31. Holland, 1998Experience with Operations Involving NORM in the UK and some Other Regions.Australian Nuclear Science and Technology Organization, Lucas heights, Sydney, March 1620

32. M. S. Hamlat, S. Gjeffal, H. Kadi, 2001Assessment of radiation exposures from naturally occurring radioactive materials in the oil and gas industry.Applied Radiation and Isotopes 55141146

33. B. Heaton, J. Lambley, 1995TENORM in the oil, gas and mineral mining industryApplied Radiation and Isotopes 46577581

34. Guidelines for naturally occurring radioactive materials2002Australian Petroleum Production & Exploration Associated Ltd. Report ABN 44000292773, March 2002, Canberra

35. Godoy MJ, da Cruz RP2003226Ra and 228Ra in scale and sludge samples and their correlation with the chemical composition.J Environ Radioact 70199206

36. Gazineu MHP, de Araujo AA, Brandao YB, Hazin CA, Godoy JM2005Radioactivity concentration in liquid and solid phases of scale and sludge generated in the petroleum industry.J Environ Radioact 814754

37. Gazineu MHP, Hazin CA2008Radium and potassium-40 in solid wastes from the oil industryAppl. Radiat. Isot. 609094

38.   Testa, C. Desideri, Meli, 1994Radiation protection and radioactive scales in oil and gas productionHealth Phys 713438

39.   Kadyrzhanov KK, Tuleushev AZ, Marabaev ZN2005Radioactive components of scales at the inner surface of pipes in oil fields of Kazakhstan. J Radioanal Nucl Chem 264413416

40.   M. Omar, H. M. Ali, M. P. Abu, 2004Distribution of radium in oil and gas industry wastes from MalaysiaAppl.Radiat Isot 60779782

41.   J. Lysebo, A. Birovliev, T. Strand, 1996NORM in oil production-occupational doses and environmental aspects. In: Proc of the 11th Congress of the Nordic Radiation Protection Society, 26-30 August 1996, Reykjavik, 137

42.   Al-Saleh FS, Al-Harshan GA2008Measurements of radiation level in petroleum products and wastes in Ryad City refinery. J Environ Radioact 9910261031

43.   Scot ML1998Naturally occurring radioactive materials in non-nuclear industry. In: Proc of the 2nd Int. Symp. on the Treatment of Naturally Occurring Radioactive Materials NORM II, 10-13 November 1998, Klefeld, Germany, 163167

44.   Oil and gas production wastes.http://www.epa.gov/rpdweb00/tenorm/oilandgas.html.

45.   European Commission Externe1994Externalities of energy. 4Oil and gas. Report EUR 16524EN. ECE, Luxembourg.

46.   Steinhäusler, Zaborowski. W. Paschoa, 2000Radiological impact due to oil-and gas extraction and processing: a comparative assessment between Asia- Pacific, Europe and South America. In: Proc of the 10th IRPA Association Congress, 14-19 May 2000, Hiroshima, P-61-285, 17

47.   E. Smith, T. Fitzgibbon, S. Karp, 1995Economic impact of potential NORM regulations In: Proc of SPA/EPA Exploration and Production. Environmental Conference,27-29 March 1995, Houston, USA, 181231

48.   Bou. Firyal-Rabee, Z. Abdallah, Rana. A. Al-Zamel, Henryk. Al-Fares, Bem, 2009NUKLEONIKA,Review paper;54(1):3−9

49.   Jonkers, F. A. Hartog, A. A. I. Knappen, P. F. J. Lance, 1997Characterization of NORM in the oil and gas production (E&P) industry. In: Proc of the NORM I, Amsterdam, 2347

50. Fischer SR1998Geologic and geochemical controls on naturally occurring radioactive materials (NORM) in produced water from oil, gas, and geothermal operations. Environ Geosciences 5139159

51. Snavely ES1989Radionuclides in produced water.Report to the American Petroleum Institute. Publication 5404API, Washington, DC, 186

52. White GJ1992Naturally Occurring Radioactive Materials (NORM) in oil and gas industry, equipment and wastes: a literature review. Report DOE/ID/01570T158.Bartlesville.

53. J. S. F. Vegueria, J. M. Godoy, N. Miekeley, 2002Environmental impact studies of barium and radium discharges by produced waters from the "Bacia deCampos" oil field offshore platforms, Brazil. J Environ Radioact 622338

54. T. Strand, I. Lysebo, 1998NORM in oil production activity levels and occupational doses. In: Proc of the 2nd Int Symp on the Treatment of Naturally Occurring Radioactive Materials NORM II, 10-13 November 1998, Klefeld, Germany, 137141

55. Norwegian Radiation Protection Authority2005Natural radioactivity in produced water from the Norwegian oil and gas industry in 2003. Report 2NRPA, Østeras.

56. Sidhu. R. Eriksen, E. Strälberg, 2006Radionuclides in produced water from Norwegian oil and gas installations-concentrations and bioavailability. CzechoslovakJ Phys 56:D43D48.

57. Al-Masri MS2006Spatial and monthly variations of radium isotopes in produced water during oil productionAppl Radiat Isot 64615623

58. United Kingdom Off-Shore Operations Association1992UK North Sea oil and gas industry; environmental inputs, impacts and issues. A report prepared by Environmental and Resource Technology Ltd, London.

59. Stephenson MT, Supernow IR1990Offshore Operators Committee 44 Platform study radionuclide analysis results. Offshore Operation Committee Report, New Orleans, Louisiana.

60. C. Swan, J. Matthews, R. Ericksen, J. Kuszmaul, 2004Evaluation of radionuclides of uranium, thorium, and radium associated with produced water fluids, precipitates and sludge from oil, gas

and oilfield brine injections wells in Mississippi. US DOE Report; DE-FG2602NT 15227.

61. Zieliński RA, Budahn JR2007Mode of occurrence and environmental mobility of oil-field radioactive material at US Geological Survey research site B. Appl Geochem 2221252137

62. W. F. Bakr, of. Assessment, radiological. the, of. impact, refining. oil, J. industry, Enviro. Radio. 101, 237 EOF243 EOF2010

63. Stephenson MT (1992) Components of produced water. J Pet Technol 548603 .

64. M. Betti, las. Aldave de, L. Heras, A. Janssens, 2004Results of the European Commission Marina II Study Part- effects of discharges of naturally occurring radioactive material.J Environ Radioact 74255277

65. Brandsma MG, Smith JP, O'Reilly JE, Ayers RC, Holmquist AL1992Modelling offshore discharges of produced water. In: Ray JP, Englehart FR (eds) Produced water. Plenum Press,New York, 5971

66. C. Hanfland, 2002Radium-226 and Radium-228 in the Atlantic Sector of the Southern OceanBer Polarforsch Meeresforssch, 43125Heaton B, Lambley JG (1995) TENORM in the oil and gas industry. Appl Radiat Isot 46:577-581.

67. Environmental, Health, and Safety Guidelines Offshore Oil and Gas Development, APRIL 30, 2007.

68. UNSCEAR,1994United Nations Committee on the Effect of Atomic Radiation: Sources and NCRP. Exposure of the population in the United States and Canada from natural background radiation. NCRP report 94national Council on Radiation Protection and Measurement, Bethesda, Maryland.

69. IAEA,2001International Atomic Energy Agency. Report of analysis on Determination of thorium and uranium naturally occurring radioisotopes in IAEA reference materials. IAEA laboratories(Chemistry Unit-01-10),Seibersdorf Austria, 15

70. E. M. El Afifi, S. A. El -Reefy, H. F. Aly, of. Treatment, waste. solid, Ra-22. containing, J. Arab, Nucl. Sci. Appl., 39200620063547

71. E. M. El Afifi, N. S. Awwad, M. A. Hilal, 2009Sequential chemical treatment of radium species in TENORM waste sludge produced

from oil and natural gas productionJournal of Hazardous Materials161907

72. M. El Afifi, M. A. Hilal, M. F. Attallah, S. A. El Reefy, of. Characterization, Wastes. Phosphogypsum, with. Associated, Acid. Phosphoric, Production. J. Fertilizers, Environ. Rad., 100, (2009407 EOF412 EOF

73. M. M. El -Didamony, N. S. Ali, M. M. Awwad, M. F. Fawzy, Treatment. Attallah, phosphogypsum. of, using. waste, organic. suitable, J. extractants, Nucl. Radioanal, D. O. I. Chem, s10967-011-1547-3,2011

Biological Monitoring of Benzene Exposure for Process Operators During Ordinary Activity in the Upstream Petroleum Industry

Magne bråtveit[1], Jorunn kirkeleit[1], Bjørg eli hollund[2], and Bente e. Moen[1]

[1]Section for Occupational Medicine, Department of Public Health and Primary Health Care, University of Bergen, Kalfarveien 31, N-5018 Bergen, Norway

[2]X lab AS, Gravdalsveien 245, N-5164 Laksevåg, Norway

ABSTRACT

This study characterized the exposure of crude oil process operators to benzene and related aromatics during ordinary activity and investigated whether the operators take up benzene at this level of exposure. We performed the study on a fixed, integrated oil and gas production facility on Norway's continental shelf. The study population included 12 operators and 9 referents. We measured personal exposure to benzene, toluene, ethylbenzene and xylene during three consecutive 12-h work shifts using organic vapour passive dosimeter badges. We sampled blood and urine before departure to the production facility (pre-shift), immediately after the work shift on Day 13 of the work period (post-shift) and immediately before the following work shift (pre-next shift). We also measured the exposure to hydrocarbons during short-term tasks by active sampling using Tenax tubes. The arithmetic mean exposure over the 3 days was 0.042 ppm for benzene (range <0.001–0.69 ppm), 0.05 ppm for toluene, 0.02 ppm for ethylbenzene and 0.03 ppm for xylene. Full-shift personal exposure was significantly higher when the process operators performed flotation work during the shift versus other tasks. Work in the flotation area was associated with short-term (6–15 min) arithmetic mean exposure to benzene of 1.06 ppm (range 0.09–2.33 ppm). The concentrations of benzene in blood and urine did not differ between operators and referents at any time point. When we adjusted for current smoking in regression analysis, benzene exposure was significantly associated with the post-shift concentration of benzene in blood ($P = 0.01$) and urine ($P = 0.03$), respectively. Although these operators perform tasks with relatively high short-term exposure to benzene, the full-shift mean exposure is low during ordinary activity. Some evidence indicates benzene uptake within this range of exposure.

INTRODUCTION

Offshore petroleum production platforms separate crude oil into oil, gas, water and solids. The water produced in this process contains a variety of chemicals, including hydrocarbons from the geological formation of the reservoir. Thus, process operators in the petroleum industry on offshore installations are potentially exposed to a mixture of hydrocarbons from crude oil, condensate from natural gas production and produced water. During ordinary operation, the processes take place in closed systems, which are opened only briefly for such purposes as sampling crude oil and the water produced, maintenance work and inspecting and cleaning pipelines and process equipment.

The upstream hydrocarbons include such aromatics as benzene, toluene, ethylbenzene and xylene. Although benzene is a known carcinogen (International Agency for Research on Cancer, 1987;Schnatter *et al.*, 2005) and a haematotoxic agent (Lan *et al.*, 2004), data on past and present exposure to benzene in the upstream petroleum industry are not extensive. Some authors (Runion, 1988; Verma *et al.*, 2000; Steinsvåg *et al.*, 2007) have summarized common occupational exposure data from oil companies obtained to document compliance with recommended limit values and have concluded that long-term mean benzene exposure is low for process operators. During routine offshore oil and gas production, full-shift mean exposure to toluene, ethylbenzene and xylene as well as to benzene is also reported to be low compared with occupational exposure limits (HSE, 1999). However, the wide ranges of exposure values indicate that some workers occasionally experience exposure exceeding the occupational exposure limit during ordinary activity (Verma *et al.*, 2000; Steinsvåg *et al.*, 2007). Previous exposure studies during ordinary activity have not provided exposure data for specific tasks expected to be associated with short-term exposure to hydrocarbons (Gardner, 2003) and have not examined whether such exposure explains some of the variability in full-shift exposure.

Full-shift benzene exposure for operators on a floating crude oil production vessel was lower during ordinary activity than during maintenance work in a cleaned crude oil tank (Kirkeleit *et al.*, 2006a). However, despite relatively low benzene exposure during maintenance work in the oil tanks (arithmetic mean 0.23 ppm), the workers had a significantly higher benzene concentration in blood and urine and acute reduction in circulating immunoglobulin M (IgM), IgA and CD4 T cells compared with referents (Kirkeleit *et al.*, 2006b,c). The internal concentration of benzene was higher than expected at the measured exposure levels, which might be related to the physically demanding 12-h work shifts for tank workers and insufficient use of respiratory protective equipment (Kirkeleit *et al.*, 2006b). Thus, benzene was absorbed and had biological effects even at exposure below the recommended limit value of 0.6 ppm and possibly within the range of exposure levels representative for ordinary activity.

This study characterized the exposure of crude oil process operators to benzene and related aromatics during ordinary activity and investigated the relationship between the individual concentrations of benzene in the breathing zone and the concentrations of unmetabolized benzene in blood and urine in these workers

METHODS

Study Site

We performed the study in October 2005 during ordinary activity in the processing unit on a fixed, integrated oil and gas production facility on Norway's continental shelf. The process operators worked 12-h shifts and surveyed the upstream processes comprising a closed system in which the effluent was separated into gas, oil, water and solid waste. They carried out this surveillance via computers in the control room and by inspection in the process areas. They also had practical tasks such as sampling and analysing oil and the water produced, fault-finding and repairing.

Sampling Strategy

Study Population

The study population originally included 10 process operators (seven men and three women) potentially exposed to benzene in the processing area and 9 referents (six men and three women) not expected to be exposed to benzene above the background concentration in the indoor environment. During the study period, two process workers were unexpectedly relocated from the cellar deck, where exposure was expected to be relatively high. To account for this relocation, we enrolled two new process operators in the study during the study period. Thus, the total number of process operators was 12. Seven referents were recruited from the catering section on the same facility, and two referents were process operators located in the central control room. We invited all eligible process workers on the selected day or night shifts and catering personnel with a shift schedule matching that of the process operators to participate. All agreed to participate.

The participants completed a self-administered questionnaire including questions on age, sex and whether they were smokers during the study period. In addition, the process operators maintained a logbook where they recorded their job tasks and use of personal protective equipment during the respective shifts.

We obtained informed written consent from all participants. We informed all subjects about their own results. The Western Norway Regional Committee for Medical Research Ethics and the Norwegian Social Science Data Services approved the study protocol. The Ministry of Health and Care Services gave permission to establish a biobank and to transfer the biological material abroad for analysis.

Personal Full-Shift Exposure to Airborne Hydrocarbons

We monitored the process operators ($n = 12$) for personal exposure to benzene, toluene, ethylbenzene and xylene during three consecutive day or night shifts (Days 11, 12 and 13 of the 2-week offshore period) each 12 h (0700–1900 or 1900–0700) using organic vapour passive dosimeter badges (3M™ 3500) attached to the worker's collar. Prior to laboratory analysis, we visually inspected the sampling badges and rejected one dosimeter due to splashes with oil-contaminated water. The arithmetic mean sampling time for the remaining 35 measurements was 657 min (range 450–730). We did not measure the personal exposure to benzene for referents.

Biological Monitoring

The original 10 process operators monitored also provided three samples of blood and urine for analysis of benzene. We collected the first sample in the morning at the heliport before departure to the oil production facility (pre-shift) and considered this the baseline measurement. We collected the second sample immediately after the work shift on Day 13 of the offshore work period (post-shift) and a third sample immediately before the work shift on the following day or night (pre-next shift). We only obtained two urine samples (post-shift and pre-next shift) from the two process operators enrolled during the study period. We obtained blood and urine samples from the nine referents on the same days following the same protocol. Due to unsuccessful blood draw, we failed to collect the pre-next shift blood sample from one of the referents. Except for the two process operators enrolled in the study later, all the participants provided urine samples pre-shift, post-shift and pre-next shift.

Exposure Measurement during Specific Work Tasks

In separate exposure measurement, we performed personal sampling of hydrocarbons during specific tasks using Tenax tubes connected to pumps (Casella-EEx) at a flow rate of 0.05–0.1 l min^{-1}. The Tenax tubes were attached to the worker's collar, and the sampling time (arithmetic mean 16.4 min, range 3–76 min) depended on the duration of the task.

In preliminary exposure assessment, we had identified four work tasks expected to be associated with relatively high short-term exposure to hydrocarbons: (i) inspection and work on the flotation package, (ii) sampling and analysis of crude oil, condensate and produced water, (iii) sending and receiving pipeline cleaning pigs and (iv) jetting of separators. The operators informed the researcher before performing these work tasks during the study period to facilitate short-term exposure measurement. Operators did not jet separators during this period.

Inspection and Work on the Flotation Package

The water produced containing dispersed oil was treated in the flotation package, where oil was skimmed off the upper layer of the two-phase water–oil mixture. Under normal conditions, the process operators do not have to inspect the flotation package regularly. During the study period, the operators inspected the flotation package twice per shift since the content of oil in the water produced exceeded the legal limit set by Norway's authorities for discharge to the sea. The operators opened the trap doors and, when necessary, adjusted the separation level. At times they also used a swab to push the oil phase over the separation edge.

Sampling and Analysis of Petroleum Streams (Crude Oil, Condensate and Water Produced)

During the night shift, operators sampled crude oil through a short-cut loop and brought it to the laboratory for analysis of water content and specific weight. During the day shift, operators manually sampled condensate and water produced in small bottles from taps in the production process and then analyzed the oil concentration of the water produced in the laboratory.

Sending and Receiving Pipeline Cleaning Pigs

The operators used a pipeline cleaning pig to remove solid or semisolid deposits or debris from the walls of the pipelines. The operators normally sent or received a pig at least once every third night shift. When inserting and launching the cleaning pig, the operators depressurized the launcher, opened the trap door and installed the pig into the launcher before locking and securing the door. When the operators received the cleaning pig from other installations, they normally left the pig within the lock for ~24 h to reduce potential hydrocarbon exposure due to evaporation caused by high temperature in the pipeline. When the operators were ready to remove the pig from the receiver, they depressurized the system and opened the trap closure. They manually pulled the pig out of the receiver and removed thick oil and wax from the pig by manual shovelling. Before they closed the trap, they cleaned and lubricated the trap closure seal.

Other Tasks with Possible Short-Term Exposure

Before operators did maintenance work on the processing equipment, they had to open, change and close blind flanges and valves. This could imply risk hydrocarbon exposure. Although process operators occasionally open process equipment, the mechanics mainly perform these tasks.

Laboratory Analysis

We stored the dosimeter badges and the Tenax tubes at −4°C before transporting them to X-lab AS in Bergen for analysis. X-lab AS desorbed the collected hydrocarbons on the dosimeter badges using CS_2 and desorbed the Tenax tubes thermically. Benzene, toluene, ethylbenzene and xylene (all isomers) were analysed quantitatively and qualitatively by gas chromatography with mass spectrometry (NIOSH, 2003). The levels of detection were 0.001 ppm for benzene and 0.01 for toluene, ethylbenzene and xylene.

Kirkeleit et al. (2006b) described methods for sampling, storing, transporting and analysing blood and urine samples in detail. We analysed the concentration of benzene in blood using a headspace sampler (Perkin Elmer Headspace sampler HS40) and a gas chromatograph (Perkin Elmer Autosystem Gas Chromatograph) using photoionization detection according to the method described by Pekariet al. (1989, 1992). We analysed the urinary concentration of benzene using a solid-phase microextraction–gas chromatograph–mass spectrometer/ion-trap detection method (SPME-GC-MS/ITD method). The limit of quantification for benzene in both blood and urine was 1 nmol l^{-1}.

Occupational Exposure Limits

Norway's recommended occupational exposure limits are averaged over an 8-h workday. In the guidelines to the Activity Regulations, the Petroleum Safety Authority Norway (2006) recommends a safety factor of 0.6 to correct for a 12-h shift, which is relevant for the offshore installations. Thus, the recommended occupational exposure limits corrected for 12-h shifts are 0.6 ppm for benzene, 15 ppm for toluene and xylene and 3 ppm for ethylbenzene. The short-term occupational exposure limits (for periods up to 15 min) are 3 ppm for benzene, 37.5 ppm for toluene and xylene and 10 ppm for ethylbenzene.

Statistical Analysis

We display personal exposure to benzene, toluene, ethylbenzene and xylene as arithmetic mean [standard deviation (SD)], geometric mean and range (minimum and maximum). We grouped the results from full-shift exposure measurement according to the logbook recordings of work tasks expected to be associated with short-term exposure to aromatic hydrocarbons. We analysed the differences in the various exposure measures between work tasks and between the process operators and referents using t-tests. We calculated Pearson's correlation coefficient for studying relationships between the different exposure measures at the various time points. We adjusted these associations for gender, age and being a current smoker by including each of these in separate multiple regression analyses. We also adjusted the associations between personal benzene exposure in the work environment and the benzene concentration in biological media for the corresponding baseline concentrations of benzene in blood and urine, respectively, by including the baseline concentration as a covariate in regression analysis.

All the exposure measurements and the biomarkers of benzene exposure had a skewed distribution, and we therefore log transformed them before statistical analysis (ln). We replaced blood and urinary benzene concentrations and benzene exposure below the limit of quantification and benzene exposure for referents by values equal to the limit of quantification divided by 2 (Hornung and Reed, 1990). We analysed the data using SPSS version 14.0.1 for Windows.

RESULTS

General Characteristics of the Study Population

The mean age (SD) of the process operators ($n = 12$) and referents ($n = 9$) was 42.3 (SD 12.8) and 44.9 (SD 10.7) years, respectively. Three process operators and four referents reported being current smokers. During the study period, the operators always used personal half-mask respirators with filter for organic vapour (brown) during work on the flotation package but not during other work tasks.

Personal Full-Shift Exposure to Benzene and Related Aromatics

Process operators' arithmetic mean benzene exposure over the 3-day study period was 0.042 ppm (range <0.001–0.69 ppm) (Table 1), which is ~7% of Norway's occupational exposure limit of 0.6 ppm over a 12-h work shift. The arithmetic mean exposure levels for toluene (0.05 ppm), ethylbenzene (0.02 ppm) and xylene (0.03 ppm) were even lower in relation to the corresponding occupational exposure limits (Table 1). The exposure to benzene was correlated with exposure to toluene ($r = 0.72$, $P < 0.001$), ethylbenzene ($r = 0.41$, $P = 0.02$) and xylene ($r = 0.70$, $P < 0.001$), respectively.

Table 1: Full-shift personal exposure to benzene, toluene, ethylbenzene and xylene for offshore process operators grouped by work task carried out during the shift

Work task	Numbers of sample (workers)	Benzene (ppm)			Toulene (ppm)			Ethylbenzene (ppm)			Xylene (ppm)		
		AM (SD)	GM	Range	AM (SD)	GM	Range	AM (SD)	GM	Range	AM (SD)	GM	Range
Flotation work	6 (2)	0.221 (0.267)	0.114	0.0030–0.068	0.21 (0.22)	0.21	0.03–0.41	0.04 (0.03)	0.03	0.01–0.09	0.13 (0.13)	0.09	0.02–0.37
Sampling	11 (4)	0.005 (0.005)	0.003	<0.001–0.014	0.01 (0.01)	0.01	<0.01–0.56	0.02 (0.01)	0.02	0.01–0.03	0.01 (0.002)	0.01	<0.01–0.03
Other work tasks	18 (6)	0.005 (0.01)	0.003	<0.001–0.023	0.01 (0.01)	0.01	<0.01–0.04	0.02 (0.01)	0.02	0.01–0.03	0.03 (0.07)	0.01	<0.01–0.01
All	35 (12)	0.042 (0.132)	0.005	<0.001–0.688	0.05 (0.11)	0.01	<0.01–0.56	0.02 (0.01)	0.02	0.01–0.09	15	0.01	<0.01–0.37
OEL (12 h)[a]		0.6			15								

Exposure did not differ significantly between the three consecutive sampling days. For benzene exposure, the arithmetic mean for the first day of sampling was 0.07 ppm (range <0.001–0.69 ppm), for the second day 0.04 ppm (<0.001–0.40 ppm) and for the third day 0.02 ppm (<0.001–0.09 ppm).

Personal exposure to benzene, toluene, ethylbenzene and xylene was significantly higher when flotation work was done during the shift compared with sampling of crude oil and water produced ($P < 0.001$) or carrying out other tasks ($P < 0.001$) (Table 1). Exposure levels did not differ when the work shift included sampling of crude oil or produced water compared with other tasks apart from flotation work. Analogous results were found for exposure to toluene, ethylbenzene and xylene (Table 1).

Benzene Concentration in Blood and Urine

The concentration of benzene in blood or urine did not differ significantly between the process operators and referents at any time point (Table 2). This result was found also among the non-smokers. The arithmetic mean concentration of benzene in blood post-shift was 1.8 nmol l^{-1} for both groups. The arithmetic mean concentration of benzene in urine post-shift was 3.9 nmol l^{-1} for process operators versus 1.6 nmol l^{-1} for referents.

Table 2: Mean concentration of benzene in blood (nmol l^{-1}) and benzene in urine (nmol l^{-1}) pre-shift, post-shift and pre-next shift for offshore process operators and referents

Marker of benzene exposure	Group	n	AN	GM	Range (minimum-maximum)	LOC
Blood, pre-shift	Process operators	10	2.1	1.4	0.5-6.0	3
	Referents	9	2.5	2.0	0.5-5.0	1

Blood, post-shift	Process operators	10	1.8	1.5	1.0-4.0	0
	Raferarts	9	1.8	1.5	1.0-4.0	0
Blood. pre-next shift	Process operators Referents	10	2.0	1.6	1.0-5.0	0
		8	1.6	1.5	1.0-3.0	0
Urine. pre-shift	Process operators Referents	10	5.8	3.0	0.5-29.0	1
		9	7.7	3.9	0.5-22.0	1
Urine, post-shift	Process operators Referents	12	3.9	1.1	0.5-34.0	7
		9	1.6	1.1	0.5-4.0	5
Urine, pre-next shift	Process operators Referents	12	1.7	1.1	0.5-9.0	4
		9	7.9	2.2	0.5-35.0	5

When we studied the associations between airborne benzene exposure and the concentration of benzene in blood and urine, we assigned referents exposure values equal to the level of detection for benzene in air divided by 2. The benzene exposure on the third day of sampling was not significantly correlated with the internal concentration of benzene immediately after work hours (post-shift) or prior to the following work shift (pre-next shift), neither in blood (post-shift: $r = 0.05$, $P = 0.83$; pre-next shift: $r = -0.08$, $P = 0.77$) nor in urine (post-shift: $r = 0.04$, $P = 0.87$; pre-next shift: $r = -0.19$, $P = 0.43$). When we included only non-smokers in the correlation analysis, benzene exposure on the third day of sampling was correlated with the post-shift concentration of benzene in blood ($r = 0.65$, $P = 0.02$). When we adjusted for being a current smoker in multiple linear regression, benzene exposure on the third day of sampling was significantly associated with the post-shift concentrations of benzene in blood ($P = 0.01$) and urine ($P = 0.03$), respectively. These models explained 87 and 79% of the variance in benzene concentration in blood and urine, respectively. Adjusting for other potential confounders such as age, gender and baseline concentrations of benzene in blood or urine did not substantially change the associations between benzene exposure and internal post-shift or pre-next shift benzene concentrations.

Short-Term Exposure during Specific Work Tasks

Work on the flotation package was associated with a short-term arithmetic mean exposure to benzene of 1.06 ppm (range 0.09–2.33 ppm), which is ~35% of Norway's recommended short-term occupational exposure limit of 3.0 ppm for periods up to 15 min (Table 3). We measured non-statistically significantly lower concentrations of benzene during work with pipeline cleaning pigs (arithmetic mean 0.32 ppm) and when opening process equipment (arithmetic mean 0.24 ppm). The apparent variability in short-term benzene exposure was higher during flotation work than when other short-term tasks were done (Fig. 1). The arithmetic mean sampling time for work in the flotation package was 9 min (range 6–15 min), for sampling of petroleum streams 19 min (3–40 min), for pipeline cleaning operations 28 min (4–76 min) and for opening of process equipment 12 min (8–16 min).

Work task	Number of operators	Benzene (ppm)			Toluene (ppm)			Ethylbenzene (ppm)			Xylene (ppm) AM (SD) GM		Range
		AM	GM	Range	AM	GM	Range	AM	GM	Range			
Work on the flotation	10	1.056	0.771	0.092-	1.06	0.82	0.19-	0.49	0.36	0.02-	1.52	1.18	0.15-
package		(0.690)		2.326	(0.69)		2.14	(0.31)		1.11	(0.97)		3.67
Sampling of petroleum	7	0.021	0.015	0.002-	0.02	0.01	<0.01-	<0.01	0.003	<0.01-	0.01	0.01	X0.01-
streams		(0.014)		0.041	(0.02)		0.04	(0.01)		0.02	(0.01)		0.03

Pipeline cleaning	6	0.322	0.258	0.088-	0.29	0.26	0.11-	0.04	0.04	0.02-	0.16	0.15	0.09-
operations		(0.220)		0.673	(0.16)		0.52	(0.02)		0.06	(0.05)		0.24
Opening of process	4	0.237	0.166	0.062-	0.19	0.08	0.02-	0.03	0.005	<0.01-	0.12	0.02	X0.01-
equipment		(0.219)		0.537	(0.29)		0.63	(0.05)		0.11	(0.21)		0.43
OEL (15 min)a		3			37.5			10			37.5		

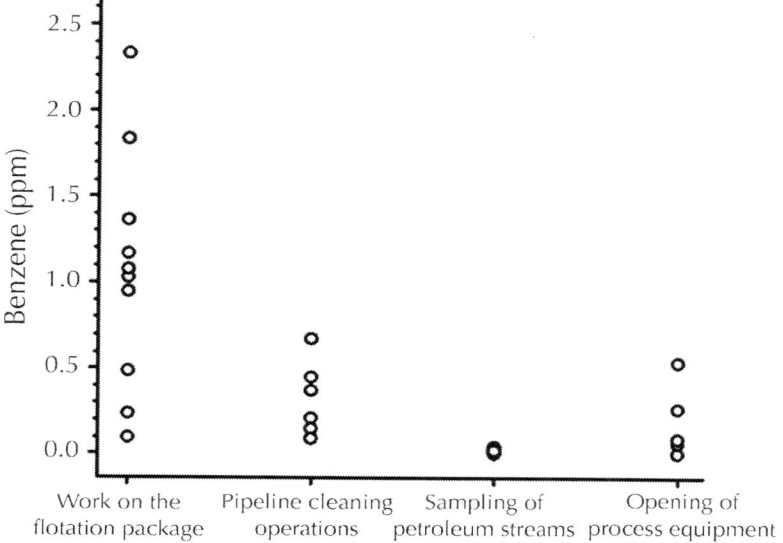

Fig. 1: Short-term personal benzene exposure grouped by work tasks on an oil and gas production facility.

Benzene exposure during sampling of crude oil, condensate and produced water (0.021 ppm) was significantly lower and had low variability compared with other short-term tasks (Table 3, Fig. 1). Differences in exposure between the different work tasks followed a similar pattern for toluene, ethylbenzene and xylene (Table 3).

DISCUSSION

During ordinary activity for offshore process operators, full-shift exposure to benzene and related aromatic hydrocarbons was generally low compared with the recommended occupational limits. Work in the flotation area was associated with relatively high short-term exposure to benzene, which contributed considerably to the full-shift exposure. Although the internal concentration of benzene did not differ between process operators and referents, benzene uptake was indicated within the range of exposure representative for ordinary activity on the installation.

In this study, we ascribe the low mean full-shift exposure to benzene at the fixed oil- and gas-producing installation to the closed process systems, which during ordinary activity are opened only briefly. In addition, the processing systems are located in open air, efficiently diluting emissions from process systems. Our low reported benzene exposure is in accordance with previous studies during ordinary activity in the offshore petroleum industry (Glass et al., 2000; Kirkeleit et al., 2006a;Steinsvåg et al., 2007). On a floating production vessel, the arithmetic and geometric mean benzene exposure levels during ordinary activity of the workers were 0.02 and 0.004 ppm, respectively (Kirkeleit et al., 2006a).Steinsvåg et al. (2007) pooled 367 personal full-shift measurements of benzene exposure in processing and drilling areas from 12 installations on Norway's continental shelf from 1994 to 2003. The benzene exposure ranged from below the limit of detection to 2.6 ppm, with arithmetic and geometric means of 0.037 and 0.0067 ppm, respectively. In retrospective exposure assessment of benzene in Australia's petroleum industry, Glasset al. (2000) estimated exposure of 0.02 ppm for workers classified as 'upstream operator offshore'. In the conventional oil and gas sector of the upstream petroleum industry in Canada, 198 personal long-term samples from 1985 to 1996 were within the range of <0.001–2.43 ppm, with an arithmetic mean of 0.064 ppm and a geometric mean of 0.011 ppm (Verma et al., 2000).

In our study, the six full-shift samples with the highest benzene exposure included brief work in the flotation area carried out by two of the 12 process operators. The supplementary short-term measurements indicated high exposure variability for benzene during this task, probably due to many factors such as the actual work done, the position of the operator, the time trap doors were open and the temperature and composition of the oil and water mixture. Several production platforms constructed during the first period of oil production on Norway's continental shelf used similar flotation package systems, but at present only a few installations have these. Sampling and analysis of crude oil and water produced did not appear to contribute significantly to full-shift exposure, and short-term measurements indicated low exposure within a narrow range for this task. On this installation, the partly automated system for sampling crude oil probably contributed to the low exposure during this task. On most offshore installations, crude oil is mainly sampled manually, which may result in higher exposure than we found. Sending and receiving a cleaning pig was associated with higher short-term exposure. However, this task was not performed during the three consecutive days of full-shift sampling. The exposure to toluene, ethylbenzene and xylene was relatively lower compared with the respective recommended occupational limit values than for benzene, indicating low health risk for these hydrocarbons.

Smoking is a major respiratory source of benzene uptake (Darrall et al., 1998). Although smoking was prohibited in the oil and gas production areas, the living quarters had a few designated smoking rooms. In our study, the internal benzene concentration did not differ between process operators and referents, but benzene in the breathing zone was associated with benzene in blood and urine, respectively, when adjusted for current smoking. Our results indicate that, even at the relatively low exposure levels during ordinary activity, workers seem to take up benzene from the working environment. Since the elimination of internal benzene is time dependent, the time lag between short-term exposure and biological sampling at the end of the work shift might have contributed to reducing the impact of such exposure on benzene uptake.

Our study included relatively few workers, and recruiting more workers would have strengthened the indicated association between benzene exposure and the internal concentration of benzene in blood and urine. Our results with strongly overlapping ranges of benzene concentration in biological media between exposed workers and referents seem to depend on both being a current smoker and exposure to benzene in the working environment. Kirkeleit et al. (2006b) showed a significant relationship, independent of smoking, between benzene exposure and internal benzene for crude oil tank workers at a geometric mean of 0.15 ppm (range 0.01–0.62), which is three times higher than our current results. The arithmetic means of the post-shift concentration of benzene in blood (17.3 nmol l^{-1}) and urine (59.3 nmol l^{-1}) of these tank workers were also considerably higher than for process operators in our present study. However, tank work should be considered a specific task, quite different from ordinary process work offshore. The concentrations of benzene measured in blood and urine in our study were in the lower part of the exposure range reported in benzene-exposed tank workers experiencing acute reduction in the circulating immune parameters IgM, IgA and CD4 T cells (Kirkeleit et al., 2006c), indicating a low risk of biological effect.

In conclusion, although offshore process operators have relatively high short-term exposure to benzene during ordinary activity, the full-shift mean exposure is low. Some evidence indicates benzene uptake within this range of exposure.

REFERENCES

1.   Darrall KG, Figgins JA, Brown RD, et al. Determination of benzene and associated volatile compounds in mainstream cigarette smoke. Analyst 1998;123:1095-101.

2.   Gardner R. Overview and characteristics of some occupational exposures and health risks on offshore oil and gas installations. Ann Occup Hyg 2003;47:201-10.

3.   Glass DC, Adams GG, Manuell RW, et al. Retrospective exposure assessment for benzene in the Australian petroleum industry. Ann Occup Hyg 2000;44:301-20.

4.   HSE. Occupational exposure to benzene, toluene, xylene and ethylbenzene during routine offshore oil and gas production. London: Health and Safety Executive; 1999. HSE Offshore Technology Report OTO 1999 088.

5.   Hornung RW, Reed LD. Estimation of average concentration in the presence of non-detectable values. Appl Occup Environ Hyg 1990;5:46-51.

6.   International Agency for Research on Cancer. Benzene. IARC Monogr Eval Carcinog Risks Hum 1987. Suppl. 7. Available at http://monographs.iarc.fr/ENG/Monographs/suppl7/suppl7.pdf. Accessed 15 June 2007.

7.   Kirkeleit J, Riise T, Bråtveit M, et al. Benzene exposure on a crude oil production vessel. Ann Occup Hyg 2006a;50:123-9.

8.   Kirkeleit J, Riise T, Bråtveit M, et al. Biological monitoring of benzene during maintenance work in crude oil cargo tanks. Chem Biol Interact 2006b;164:60-7.

9.   Kirkeleit J, Ulvestad E, Riise T, et al. Acute suppression of serum IgM and IgA in tank workers exposed to benzene. Scand J Immunol 2006c;64:690-8.

10.   Lan Q, Zhang LG, Li G, et al. Hematotoxicity in workers exposed to low levels of benzene. Science 2004;306:1774-6.

11. NIOSH. NIOSH manual of analytical methods. 3rd edn. Cincinnati, OH: National Institute of Occupational Safety and Health; 2003. (Hydrocarbons, Method: 1501).

12. Pekari K, Riekkola M-L, Aitio A. Simultaneous determination of benzene and toluene in the blood using head-space gas chromatography. J Chromatogr 1989;491:309-20.

13. Pekari K, Vainiotalo S, Heikkilä P, et al. Biological monitoring of occupational exposure to low levels of benzene. Scand J Work Environ Health 1992;18:317-22.

14. PSAN. Guidelines to regulations relating to conduct of activities in the petroleum activities (the Activities Regulations): section 34. Stavanger, Norway: Petroleum Safety Authority Norway; 2006.

15. Runion HE. Occupational exposure to potentially hazardous agents in the petroleum industry. Occup Med 1988;3:431-44.

16. Schnatter AR, Rosamilia K, Wojcik NC. Review of the literature on benzene exposure and leukemia subtypes. Chem Biol Interact 2005;153–154:9-21.

17. Steinsvåg K, Bråtveit M, Moen BE. Exposure to carcinogens for defined job categories in Norway's offshore petroleum industry, 1970–2005. Occup Environ Med 2007;64:250-8.

18. Verma DK, Johnson DM, McLean JD. Benzene and total hydrocarbon exposures in the upstream petroleum oil and gas industry. Am Ind Hyg Assoc J 2000;61:255-63

Electromagnetic Heating of Heavy Oil and Bitumen: A Review of Experimental Studies and Field Applications

Albina Mukhametshina[1,2] and Elena Martynova[1,2]

1Harold Vance Department of Petroleum Engineering, Texas A&M University, 3116 TAMU-407 Richardson Building, College Station, TX, USA 2Gubkin Russian State University of Oil and Gas, 65 Leninsky Prospekt, Moscow, Russia

ABSTRACT

Viscosity is a major obstacle in the recovery of low API gravity oil resources from heavy oil and bitumen reservoirs. While thermal recovery is usually considered the most effective method for lowering viscosity, for some reservoirs introducing heat with commonly implemented thermal methods is not recommended. For these types of reservoirs, electromagnetic heating is the recommended solution. Electromagnetic heating targets part of the reservoir instead of heating the bulk of the reservoir, which means that the targeted area can be heated up more effectively and with lower heat losses than with other thermal methods. Electromagnetic heating is still relatively new and is not widely used as an alternate or addition to traditional thermal recovery methods. However, studies are being conducted and new technologies proposed that could help increase its use. Therefore, the objective of this study is to investigate the recovery of heavy oil and bitumen reservoirs by electromagnetic heating through the review of existing laboratory studies and field trials.

INTRODUCTION

High-frequency electromagnetic radiation is a relatively new technique for use in enhanced oil recovery methods. It has been tested by theoretic, laboratories and field trial research in Russia [1–10], the United States [11–17], Canada [18–21], and other countries [22–34]. Traditional thermal recovery and well stimulation techniques using hot steam or fluid are not effective in some cases [7, 35] due to prohibitive heat losses from injection wells and reservoirs, low reservoir injectivity (especially for bitumen deposits), steam leakage, large overburden heat loss at thin pay zones, permafrost conditions, and so forth. Furthermore, commonly used thermal recovery methods are not considered environmentally friendly, damaging the hydrogeologic environment and contributing to the greenhouse effect.

The most important thing in electromagnetic heating is that the heat is developed within the material rather than being brought from outside, which means the material is heated more uniformly throughout the medium [27]. Therefore, instead of heating the bulk reservoir volume, part of the reservoir can be targeted and heated more effectively with lower heat loss than other thermal methods. Unlike traditional thermal recovery methods, microwave heating causes friction by vibration of molecules, which results in dielectric heating of the reservoir. Heat and mass transfer in different environments under microwave influence was studied by a number of scientists around the globe, but its application as an EOR method is not yet fully understood. Microwave heating is not used productively because of the lack of reliable information about the processes of heat and mass transfer in a multiphase system in porous media under the influence of electromagnetic radiation, which does not allow effective control. Therefore, current research studies use modeling to discover optimum design parameters for the use of microwave heating in field applications.

Review of Experimental Electromagnetic Heat Studies

The success of near wellbore heating with electromagnetic waves has been proven experimentally [1]. To represent reservoir rock, quartz sand with a 7.7 dielectric constant and a 0.083 tangent was used in the laboratory. A 20% initial water saturation and an 80% initial oil saturation with 16.61 cps (Pa·s) gravity and 0.86 g/cm^3 (860 kg/ m^3) density oil were maintained. The dielectric constant and the loss tangent of the oil sample were 2.23 and 0.019, respectively. In this study of electric and high-frequency electromagnetic heating of a reservoir model, the temperature of the medium was measured by thermometers located at different points of the experimental setup. In another case, a linear radiator with a length of 87 cm and a diameter of 19 mm was placed in the center of the setup. The linear radiator was connected via acoaxial cable to the generator supplying electromagnetic waves with a frequency of 13.56 MHz. In the experiment, it was discovered that when fluid temperature was exposed to a high-frequency electromagnetic field at the same distance from the radiation source it was greater than one initiated by electric heating. In this case, the thermal conductivity of the medium is affected only slightly. When heat is induced throughout the whole volume of the medium, the amount of heat introduced depends largely on the electrical properties of the medium. The study summarized the advantages of microwave heating over electrical heating as having deeper penetration, quicker heating, and lower heat losses.

Sayakhov [2] discussed the physical foundations of fluid filtration in high-frequency fields. Radial fluid filtration through porous media in a heterogeneous high-frequency electromagnetic field was studied experimentally. A specially designed core holder, filled with a representative porous medium, was located in a coaxial resonator. A high-frequency wave generator was used to generate mw energy, which was directed to the resonator through cables. Studied liquid came into the porous media through the hose first, then through the lumen of the inner conductor of the resonator dripped into a graduated cylinder. Experiments were started with a 200–300 watts power microwave oven at 2400 MHz frequency and 500 watts vibrational power. Temperatures were recorded throughout the experiments via thermocouples inserted in the core holder. Kerosene was used as a representative reservoir fluid. The measurements were performed with and without the influence of a high-frequency electromagnetic field. It was established that exposure to high-frequency electromagnetic fields leads to a sharp increase in flow rate per unit of time and fluid temperature at the outlet. In addition, the flow rate increases dramatically once exposed to the electromagnetic field, while the temperature increases after 10 seconds. After the field is discontinued, a sharp decrease in the flow rate is observed and a gradual cooling of the porous medium takes place.

Experimental studies [3] on the influence of electromagnetic fields with a frequency of Hz to Hz on thermal conductivity of dielectric liquids showed that the thermal conductivity of the liquids increased when exposed. The thermal conductivity increases as the magnitude of the dipole moment of the liquid used in the experiment parameters increases with the frequency and intensity of field.

Fatikhov [5] conducted experimental research on the flow of bitumen oil at different pressure gradients in a high-frequency electromagnetic field. The experiment focused on changes in the volumetric flow rate of filtered oil under different pressures at different temperatures. The initial pressure drop for bitumen from the Mordovo-Karmalskoye deposit in reservoir conditions was 0.003 MPa/m. During the experiment, the pressure drop decreased rapidly as the temperatures were increased, resulting in bituminous oil becoming a Newtonian liquid. Therefore, it was established that the application of electromagnetic heating improves fluid flow behavior and the non-Newtonian properties of bitumen decrease rapidly with increasing temperature.

In 1992, Kasevich et al. [17] studied electromagnetic heating of rock samples at 1 kW (frequency 50.55 MHz) and 200 W (144 MHz) and found that a particular type of rock could be heated to 423 K when exposed to an RF electromagnetic field. The rock, which had low thermal conductivity, heats poorly when hot steam is pumped into the reservoir. They conducted experiments at both normal and formation pressures.

Ovalles et al. [30] used a microwave with 650 watts power to heat core samples saturated with oil of 25 API gravity and 7.7 API gravity (a sample from the Orinoco River Basin). Medium API oil temperatures were measured in 0.5, 1, and 1.5 minute intervals and heavy oil intervals were increased to 1, 5, and 10 minutes. The experimental results were used to test mathematical models and predict the production of three abstract oil reservoirs in Venezuela.

Chakma and Jha [23] conducted laboratory experiments using electromagnetic heating on a scaled thin heavy oil reservoir pay zone model. Gas injection with horizontal wells during electromagnetic heating was achieved. The aim was to decrease oil viscosity with electromagnetic heating and obtain a gas drive with the injected gas. Using nitrogen for the injected gas, they were able to prove that for thin pay zones heating of the wellbore vicinity is sufficient, by achieving oil recoveries as high as 45% of original oil in place compared to estimated primary recovery rates of less than 5%. Recovery achieved by use of the combined method was higher than that of nitrogen injection or electromagnetic heating alone. Chakma and Jha also discussed a number of parameters affecting the results of the combined method, including

- gas injection pressure (when no gas was injected, oil was produced only due to gravity drainage and no significant convective transport occurred; therefore, gas injection provided an oil rate increase with the increasing injection pressure);
- temperature (the initial production rate was not significantly affected by temperature, but later there was an increase in the production rate, meaning that overall recovery increased with temperature);

- electromagnetic frequency (the higher the frequency, the greater the recovery);

- oil viscosity (as expected, a higher oil viscosity leads to a lower recovery for a given electromagnetic frequency, temperature, and gas injection pressure);

- salinity (higher salinity provides higher recovery due to the higher conductivity of saline water compared with distilled water);

- electrode distance (recovery is similar, but closer electrode spacing provides faster production rates) [23].

Hascakir et al. [24] conducted a laboratory study of microwave-assisted gravity drainage on heavy oil samples from reservoirs in Turkey (Bati Raman, 9.5 API; Garzan, 12 API; and Camurlu, 18 API) using a specially designed novel graphite core holder packed with crushed limestone. Their study described the effects of operational parameters like heating time, waiting period and rock, and fluid properties on the effectiveness of microwave heating. Some of the conclusions made are confirming ones found in Chakma and Jha [23], like the positive effect of high water salinity and water saturation. Hascakir et al. [24] also concluded that water wet conditions are preferable for obtaining higher oil recoveries and that large porosity and permeability are also favorable. When microwave heating is applied to oil samples continuously, higher temperatures are reached, which allows better results to be achieved than with periodic heating when microwave heating is applied for a limited time in periodic intervals. Therefore, higher temperatures allow for better results in continuous heating.

Jha et al. [27] proposed using microwave-assisted gravity drainage (MWAGD) in the Mehsana oil field in India. They heated specially prepared samples with the required characteristics from that field in the laboratory using a microwave with variable power up to 1000 watts operating at 3 GHz frequency, which allowed them to obtain temperature and viscosity profiles of the gravity-drained oil. They described effects of initial oil and water saturations, wettability, porosity, and permeability similar to those found by Hascakir et al. [24] and Chakma and Jha [23].

Jha et al. [27] suggested using MWAGD commercially by drilling one horizontal well and multiple vertical ones with downhole microwave antennas. However, this might not allow deep enough heat penetration, so other options are also proposed such as a combination of two horizontal wells and installing antenna inside the horizontal production well. Vertical separation of the horizontal well pair is approximately 15 meters, which is far more efficient than SAGD in which the separation is around 5 meters. Because crude oil absorbs microwave heat weakly, Jha et al. also proposed increasing thermal conductivity by injecting powdered metallic oxides, chlorides, or activated carbon through a fracture operation. The working principle and description of laboratory applications of such additives to heavy oil can be found in various studies by Hascakir et al., Kershaw et al., and Odenbach [25, 28, 29].

Technical principles of the SAGD method assisted by electromagnetic heating (EM-SAGD process) were reported by Koolman et al. [26]. Inductive heating was initiated in the laboratory using an EM source with a working frequency of 142 kHz. The sample was heated for 10 minutes at a power of 7.2 kW, achieving a rise in the temperature of 7.5 K. Laboratory and field processes were modeled using a numerical simulator, combining electromagnetic and thermal modules. It was specially built and can be applied to field-scale simulations. According to simulation results, a 38% increase in bitumen production was predicted compared to conventional SAGD.

Kovaleva et al. investigated the effects of radio frequency electromagnetic (RF-EM) fields and electrical heating on the mass- and heat-transfer processes in a multicomponent hydrocarbon system flowing in porous media [4, 6]. Three different types of experiments were carried out: solvent (kerosene) flooding under the RF-EM field, solvent (kerosene) flooding under electrical heating, and cold solvent (kerosene) flooding. In all three experiments, the physical characteristics of the model and heating conditions (temperatures) were identically maintained. Two series of experiments on models with different granulometric composition of formation were also carried out.

Figure 1 shows the dependence of oil recovery on the volume of solvent injected. This figure demonstrates that the highest oil recovery was obtained by applying an RF-EM field. Kovaleva et al. [4, 6] concluded that following RF-EM influence on the oil-saturated samples the quantity of the received oil is more than the quantity received under electrical (and thermal) processing at identical temperatures of heating of the media. It confirms the additional "nonthermal" action of the electromagnetic field.

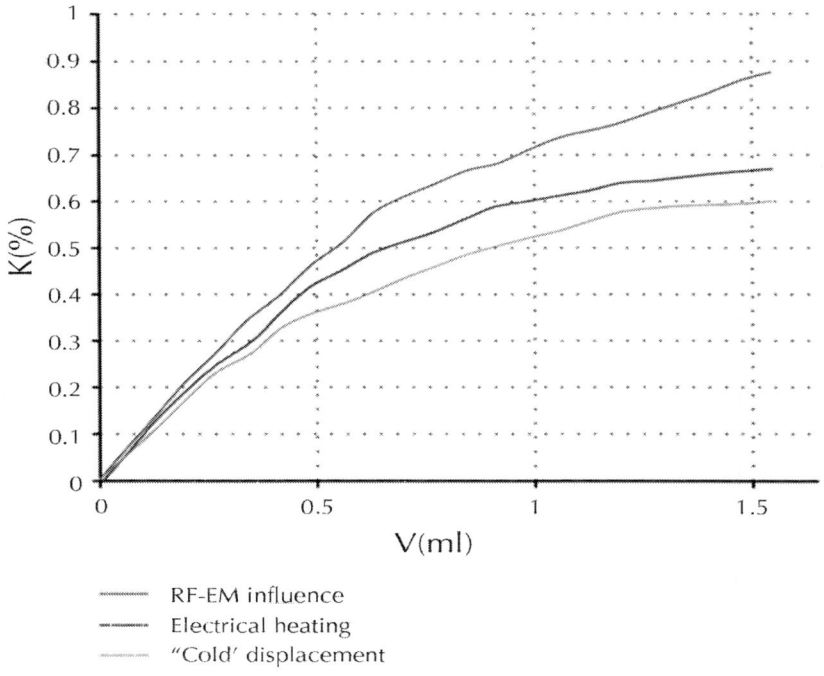

RF-EM influence
Electrical heating
"Cold' displacement

Figure 1: Dependence of oil recovery () on the relative volume of solvent injected. Adapted from—[6]

TECHNOLOGIES OF IN SITU ELECTROMAGNETIC HEATING OF HEAVY OIL AND BITUMEN

The first production method applying microwave heating to well production was patented in 1956 [14]. Electromagnetic waves were transferred to the well bottom from the surface through a coaxial system of internal and external pipes (tubing and casing). Interaction of electromagnetic waves with the formation causes the emergence of distributed volumetric heat sources and reduces the viscosity of the reservoir fluid. In 1965, Haagensen [13] described a device for generating high-frequency electromagnetic waves at the mouth of the well and a method of delivering electromagnetic energy through coaxial lines and waveguides to the bottom hole. In 1987, Wilson [16] described a similar device in his work, with some modifications of the radiating element of EM waves.

A huge drawback to the methods described by Haagensen and Wilson [13, 16] is the shallow penetration of electromagnetic waves, and, hence, a low sweep efficiency of heating. When the method described by Ritchey [14] was implemented, there were large losses of electromagnetic energy. Due to the finite conductivity of tubing, they are heated and electromagnetic energy is dissipated in rocks surrounding the well, resulting in large wellbore heat losses, especially if a permafrost layer is present.

Sayakhov et al. [8] proposed a method of recovery that included creating a combustion front with a simultaneous electromagnetic current influence. The environment is heated when exposed to an electromagnetic field, which decreases the viscosity and increases the mobility of crude oil. It is assumed in this method that the EM field continues influencing the reservoir after the combustion is initiated.

Review of Electromagnetic Heating Field Studies

Electromagnetic heating field trials have been carried out in Russia (Bashkortostan and Tatarstan) [2, 9, 10], the United States (California and Utah) [11, 17], and in Canada (Alberta and Saskatchewan) [18–21].

Russia

In Russia, field tests of radio frequency electromagnetic heating of the near-wellbore zone were first launched in 1969 at Well 40/19 in the Ishimbayskoye Oil Field in Bashkortostan and continued in the Yultimirovskoye Bitumen Field in Tatarstan according to Sayakhov et al. [2, 9, 10]. Characteristics of Well 40/19 from the Ishimbayskoye Oil Field are represented in Table 1.

Table 1: Characteristics of Well 40/19 of the Ishimbayskoye Oil Field

Parameter	Value	
Depth, m		830
Casing diameter, in.		6
Tubing diameter, in.		2
Flow rate, ton/day		3
Well temperature, K	287-289	
Paraffin content, 96		2.3
Resins content, 96		11
Density, kg/m³		890
Viscosity at 20°C, m²/c	20	* 106

The source of high-frequency electromagnetic energy was a generator providing an optimum oscillation output power of 63 kW at a frequency of 13.56 MHz. Standard RF coaxial cable was used to supply high-frequency electromagnetic energy from the generator to the well. Temperature measurements were carried out using a thermograph at the bottom hole. Temperature was recorded continuously at a fixed depth of 650–655 m (in the open hole, where the radiating element was working) while the generator was operating (Table 2).

Table 2: Dynamics of temperature growth

Heating time, days	Temperature increase at the bottom hole, K
0.5	283
1	290
2	300
3	306
4	311
5	313

Yultimirovskoye Bitumen Field

RF electromagnetic heating was conducted in the Yultimirovskoye Bitumen Field by Sayakhov in 1980 [2] (Table 3). Two wells spaced 5 meters apart were studied, Well 150 and Well 1.

Table 3: Characteristics of the Yultimirovskoye reservoir

Parameter	Value
Porosity, %	25
Bitumen saturation, 96	3.6
Permeability, znicm²	0-0.183

Electromagnetic heating of the bitumen reservoir was conducted in several stages at different conditions. Initially, the RF-EM installation was set to about of 20 kW power. After 36.5 hours, the temperature at the bottom of Well 150 has increased from 282 to 389 K. No temperature change was observed in Well 1. In the next phase, RF-EM installation was reset to approximately 30 kW of power. As a result, the temperature in Well 150 reached 423 K after six hours. It should be noted that the growth rate of temperature increased. In the third phase, the RF-EM installation was set to a maximum of 60 kW, which caused the heating intensity in Well 150 to increase greatly. After 5.5 hours, the temperature in Well 150 increased from 417 to 463 K. Next, the RF-EM unit was turned off for 2 hours, resulting in a temperature drop to 423 K.

During the next 32 hours, the RF-EM installation operated at maximum output, causing the heating of the bottom hole of Well 150 up to 583 K and the bottom hole of Well 1 up to 318 K. In Sayakhov's experiment [2], the fluoroplastic collars that centered the tubing in Well 150 melted (their maximum operating temperature was 573 K). As a result, a short circuit between the casing and tubing occurred and the RF-EM unit broke and was disabled. Before this disruption, the RF-EM installation worked steadily throughout the field experiment.

Deep heat penetration (up to 5 m in the reservoir) was demonstrated by temperature measurements done during the cooling of Well 150's drain zone after the exposure. The temperature decreased from 373 K to 343 K in three days. After a few days of electromagnetic heating, wellbore heat distribution revealed low heat losses to both the overburden and underburden.

THE UNITED STATES

Bakersfield, California, United States

In 1992, Kasevich et al. [17] conducted field tests of RF-EM heating in the United States in the Bakersfield, California, field. The goal was to prove the concept that controlled RF-EM radiation could be used as a thermal EOR method. The production of reservoir fluids was not measured because it was a quality, not quantity, study designed to gain a better understanding of underground processes.

A high-frequency electromagnetic wave generator with a capacity of 25 kW and frequency of 13.56 MHz was used to heat the reservoir at Well 100D. Heat penetration was determined by temperature measurements in the surrounding observation wells T10, T20, and T30, which were located 3, 6, and 9 meters from Well 100D. Kasevich et al. [17] also proved that the RF producer used could efficiently focus its radiation pattern into the desired region by measuring return loss and electromagnetic radiation. In well T10, situated 3 meters from Well 100D, the medium temperature was increased from 293 K to 393 K over 20 hours of EM heating.

Avintaquin Canyon and Asphalt Ridge (Utah)

In 1980, one of the most detailed studies on electromagnetic-heat-based oil recovery was done at the Illinois Institute of Technology Research Institute (IITRI) by Bridges et al. [11]. They carried out extensive research work on the use of the different types of electromagnetic heating for different types of deposits, oil shale, and tar sand.

Bridges et al. [11] tested their IITRI technique of RF electromagnetic heating with two field experiments in Avintaquin Canyon, Utah, USA. Shale that was 6 meters thick was found in outcrops convenient for relatively cheap horizontal experiments. So arrays of holes were drilled and electrodes were inserted to a depth of 1 meter. These tests allowed the researchers to gain experience and to prove it was possible to achieve in situ pyrolysis of oil shale, thus increasing its thermal maturity. The power applied to the formation ranged from 5 kW to 20 kW, with a frequency of 13.56 MHz. As a result of EM heating, temperatures rose to 673 K and 20–30 % of the oil content was collected. However, it should be noted that the amount of produced oil was badly affected by the presence of cracks which allowed light hydrocarbons to escape (evaporate).

In 1981, Bridges et al. [11] conducted field tests on tar sand at Asphalt Ridge, Utah, USA. The first experiment tested the gravity drive bitumen recovery process, designed to prove EM heating concepts and improve equipment design. This experiment used vertical electrode placement and a mined collection chamber and tunnel. It was equipped with a 200 kW radio transmitter and heated 25 m^3 of tar sand. In the first experiment, the roof of the mined chamber was not supported well enough, which resulted in early termination of the experiment. Therefore, the situation had to be fixed by constructing a concrete arch for the second pilot test. The heating power used on tar sand varied from 40 kW to 75 kW with a frequency of 13.56 MHz.

The second test quantified the results of heating over a longer period and at higher temperatures. In this experiment, temperatures exceeded 473 K, and 30 to 35% recovery was achieved in just 20 days. This was encouraging because continuation of heating could have resulted in even higher recovery. The power loss was minimal in all the experiments, which proved the efficiency of heating.

Canada

The Wildmere Field, Alberta, Canada

According to Spencer [18] commercial EM heating was first introduced in the field at Wildmere, Alberta, Canada (Table 4). The first well was drilled in January 1986 and began producing oil in March of the same year. Before EM heating began in May, the well was producing about 0.95 tonnes/day. After EM heating commenced, production rates increased and soon settled at the level of 3.18 tonnes/day until November 1986, when the well was closed due to technical reasons.

Another well in this field increased in production from 1.59 tonnes/day to an average of 4.77 tonnes/day, with the maximum flow rate reaching 9.54 tonnes/day.

Table 4: Characteristics of the reservoir in the Wildmere field, Alberta, Canada

Parameter	Value
Net thickness, m	1
Depth, m	600
Density, kg/m³	987
Oil viscosity at 20°C, Pass	20

The Lloydminster Heavy Oil Area, Saskatchewan, Canada

In 1988-1989, two electromagnetic stimulation projects were conducted by Davidson [22] in the Lloydminster heavy oil area in Saskatchewan, Canada. Unfortunately the economic potential of the process could not be evaluated from either pilot test, because long-term heating could not be achieved due to equipment failure (casing insulation) and special reservoir conditions. However, the technical results looked promising.

The first pilot well was in Northminster (Saskatchewan, Canada) and produced 11.4 API oil from the Sparky formation. The power was applied to the well in a pulsating manner with a baseline of 20 kW with four-hour spikes of 30 kW (2 daily), in order to reduce the risk of significant damage to the insulation. Later, the power was increased to a 25 kW baseline with 35 kW pulses and finally to a 30 kW baseline with 50 kW peaks. At that point the insulation failed, and the power rates came down to 28 kW. Power rates were later leveled at 47 kW and stayed that way until terminated.

As can be seen from Table 5, water cut and production both reacted positively to electromagnetic stimulation. However, it should be noted that some portion of the increased production is related to the increase in pump speed. Water cut drop can be related directly to the

EM effect and improvement in oil mobility. Once the heating has been terminated, technical parameters return rapidly to their initial states

Table 5: Oil field performance for the Northminster pilot: primary and achieved by EM heating.

Parameter	Primary EM heating	
Production rate, m³/day	10-12	20
Water cut, 96	15-20	10-12
Productivity index, bbl/psi	0.33	0.42
Stimulation ratio		1.27

The second pilot well was situated in Lashburn (Saskatchewan, Canada) and produced very viscous 11.4 API crude oil from the Sparky formation. During the reservoir heating phase, the power ranged from 13 to 18 kW. This well has high sand cuts and before electrical power was applied its production was not stable and regularly had to be stimulated by flushing the wellbore. The peak production reached 5.0 m³/day and had begun to decline before electromagnetic heating was applied. Electromagnetic heating reduced the water cut, but the well was still prone to high water cuts after shut-in periods. Oil production also increased when electromagnetic forces were applied, until it reached 9 m³/day. During the initial heating phase of the reservoir, the temperature in the bottom hole increased steadily from 295 K to 309 K, but temperatures began dropping immediately after the power was turned off or when power delivery systems failed.

ESEIEH (Alberta, Canada)

Enhanced Solvent Extraction Incorporating Electromagnetic Heating technology (ESEIEH) has been patented and is currently undergoing tests, as reported by Rassenfoss [20]. The ESEIEH consortium is relying on three oil company partners to help with this testing: Laricina Energy, Nexen, and Suncor Energy. The pilot project is planned to take three years and currently is in the first stage. Field application is expected to start later in 2013. The ESEIEH method combines the familiar horizontal well pair design commonly used in the Canadian oil sands, coupled with heating using RF-EM waves and solvents, such as butane or propane. The company aims to heat the reservoir by running an antenna underground that emits enough energy to raise the temperature to 50°C (120°F)

CONCLUSION

A review of electromagnetic heating for enhanced oil recovery was presented in this paper. A number of studies show that electromagnetic heating is a promising method of enhanced oil recovery. However, the studies to date are limited, and only a few field trials have been reported. Most of the current research is based on laboratory experiments or numerical models. It should be noted that this paper did not cover the computer simulations carried out to research the effectiveness of EM heating.

Better understanding of the in situ electromagnetic process is essential and can be achieved by combining laboratory, numerical, and field-scale tests. At the moment it is not possible to assess the efficiency of EM heating or the opportunities for economic applications of it alone or in combination with traditional methods; therefore, more global studies should be conducted.

Even though sustainability of this technology has not yet been completely evaluated, the method definitely should not be overlooked by the industry because of its enormous potential. Attempts should be made to develop viable screening criteria for possible production of heavy oil, oil shale, and tar sand deposits.

REFERENCES

1.   S. Chistyakov, F. Sayakhov, and G. Balabyan, "Experimental study of formations dielectric properties under the influence of high-frequency electromagnetic fields," in University Investigations: Geology and Exploration, pp. 153–156, 1971.

2. F. Sayakhov, "Particular properties of filtration and fluid flow under the influence of high-frequency electromagnetic field," in Joint University Scientific Book, pp. 108–120, 1980.

3. B. Savinikh, V. Dyakonov, and A. Usmanov, "The influence of alternating electric currents on the thermal conductivity of dielectric fluids," Journal of Engineering Physics and Thermophysics, no. 2, pp. 269–276, 1981 (Russian).

4. A. Davletbaev and L. Kovaleva, "Combined RF EM/solvent treatment technique: heavy/extra-heavy oil production model case study," in Proceedings of the 10th Annual International Conference Petroleum Phase Behavior and Fouling, Rio de Janeiro, Brazil, 2009.

5. M. A. Fatikhov, "Experimental study of bitumen initial pressure gradient in the electromagnetic field,"University Investigations: Oil and Gas, no. 5, pp. 93–94, 1990 (Russian).

6. L. Kovaleva, A. Davletbaev, T. Babadagli, and Z. Stepanova, "Effects of electrical and radio-frequency electromagnetic heating on the mass-transfer process during miscible injection for heavy-oil recovery,"Energy and Fuels, vol. 25, no. 2, pp. 482–486, 2011.

7. G. Malofeev, O. Mirsaetov, and I. Cholovskaya, "Injection of hot fluids for enhanced oil recovery and well stimulation," in Regular and Chaotic Dynamics, Institute of Computer Science, Russialgevsk, Russia, 2008.

8. F. Sayakhov, R. Bulgakov, V. Dyblenko, B. Deshura, and M. Bykov, "About HF heating of bitumen reservoirs," Petroleum Engineering, no. 1, pp. 5–8, 1980 (Russian).

9. F. L. Sayakhov, L. A. Kovaleva, M. A. Fatikhov, and G. A. Khalikov, "Method of thermal effect on oil-bearing formation," SU Patent 1723314, 1992.

10. F. Sayakhov, I. Habibullin, M. Yagudin, and M. Fatyhov, "Technique and technology of thermal well stimulation on the basis electro-thermo-chemical and electromagnetic effects," University Investigations: Oil and Gas, no. 2, pp. 33–42, 1992 (Russian).

11. J. E. Bridges, J. J. Krstansky, A. Taflove, and G. C. Sresty, "The IITRI in situ RF fuel recovery process,"Journal of Microwave Power, vol. 18, no. 1, pp. 3–14, 1983.

12. J. Bridges, "Method for in-situ heat processing of hydrocarbonaceous formation," US Patent 4140180, 1979.

13. A. D. Haagensen, "Oil well microwave tools," Patent USA 3170119, 1965.

14. H. W. Ritchey, "Radiation Heating System, US Patent," Tech. Rep. 2757738, 1956.

15. G. C. Sresty, R. H. Snow, and J. E. Bridges, "Recovery of liquid hydrocarbons from oil shale by electromagnetic heating in-situ," US Patent 4485869, 1984.

16. R. Wilson, "Well production method using microwave heating," US Patent 4485868, 1987.

17. R. S. Kasevich, S. L. Price, D. L. Faust, and M. F. Fontaine, "Pilot testing of a radio frequency heating system for enhanced oil recovery from diatomaceous earth," in Proceedings of the SPE Annual Technical Conference & Exhibition, pp. 105–113, New Orleans, La, USA, September 1994.

18. H. L. Spencer, "Electromagnetic Oil Recovery, Ltd," Calgary, Canada, 1987.

19. F. E. Vermeulen and F. S. Chute, "Electromagnetic techniques in the in-situ recovery of heavy oils," Journal of Microwave Power, vol. 18, no. 1, pp. 15–29, 1983.

20. S. Rassenfoss, "Seeking more oil, fewer emissions," Journal of Petroleum Technology, vol. 64, no. 9, pp. 34–38, 2012.

21. B. C. W. Mcgee and F. E. Vermeulen, "The mechanisms of electrical heating for the recovery of bitumen from oil sands," Journal of Canadian Petroleum Technology, vol. 46, no. 1, pp. 28–34, 2007.

22. R. J. Davidson, "Electromagnetic stimulation of Lloydminster heavy oil reservoirs: field test results,"Journal of Canadian Petroleum Technology, vol. 34, no. 4, pp. 15–24, 1995.

23. A. Chakma and K. N. Jha, "Heavy-oil recovery from thin pay zones by electromagnetic heating, paper SPE 24817," in Proceedings of the Annual Technical Conference and Exhibition, Society of Petroleum Engineers, Washington, DC, USA, October 1992.

24. B. Hascakir, C. Acar, Schlumberger, B. Demiral, and S. Akin, "Microwave assisted gravity drainage of heavy oils," in Proceedings

of the International Petroleum Technology Conference (IPTC '08), pp. 1908–1916, Kuala Lumpur, Malaysia, December 2008.

25. B. Hascakir, T. Babadagli, and S. Akin, "Experimental and numerical modeling of heavy-oil recovery by electrical heating, paper SPE 117669," in Proceedings of the International Thermal Operations and Heavy Oil Symposium (ITOHOS '08), p. 14, Society of Petroleum Engineers, Alberta, Canada, October 2008.

26. M. Koolman, N. Huber, D. Diehl, and B. Wacker, "Electromagnetic heating method to improve steam assisted gravity drainage, paper 1177481," in Proceedings of the International Thermal Operations and Heavy Oil Symposium (ITOHOS '08), pp. 327–338, Society of Petroleum Engineers, Alberta, Canada, October 2008.

27. K. A. Jha, N. Joshi, and A. Singh, "Applicability and assessment of micro-wave assisted gravity drainage (MWAGD) applications in Mehsana heavy oil field, paper SPE 14591," in Proceedings of the SPE Heavy Oil Conference and Exhibition, Society of Petroleum Engineers, Kuwait City, Kuwait, December 2011.

28. J. R. Kershaw, G. Barrass, and D. Gray, "Chemical nature of coal hydrogenation oils part I. The effect of catalysts," Fuel Processing Technology, vol. 3, no. 2, pp. 115–129, 1980.

29. S. Odenbach, "Ferrofluids—magnetically controlled suspensions," Colloids and Surfaces A, vol. 217, no. 1–3, pp. 171–178, 2003.

30. C. Ovalles, A. Fonseca, A. Lara et al., "Opportunities of downhole dielectric heating in Venezuela: three case studies involving medium, heavy and extra-heavy crude oil reservoirs, paper SPE 78980," inProceedings of the International Thermal Operations and Heavy Oil Symposium and International Horizontal Well Technology Conference, Alberta, Canada, November 2002.

31. M. A. Ayrapetyan, "About oil fields development prospects by high-frequency currents electrical fields," inMaterials of KSSR Institute of Oil, pp. 38–52, 1958.

32. M. A. Ayrapetyan, V. S. Velikanov, and E. Ya. Magnikov, "Reservoir high-frequency heating investigations," in Materials of KSSR Institute of Oil, pp. 113–124, 1959.

33. M. A. Carrizales, L. W. Lake, and R. T. Johns, "Production improvement of heavy-oil recovery by using electromagnetic

heating, paper SPE 115723," in Proceedings of the SPE Annual Technical Conference and Exhibition (ATCE '08), Denver, Colo, USA, September 2008.

34. A. D. Hiebert, F. E. Vermeulen, F. S. Chute, and C. E. Capjack, "Numerical simulation results for the electrical heating of Athabasca oil-sand formations," SPE Reservoir Engineering, vol. 1, no. 1, pp. 76–84, 1986.

35. J. Burge, P. Surio, and M. Combarnu, Thermal Methods of Enhanced Oil Recovery, Nedra Publishing, Moscow, Russia, 1988.

Chapter 9

Characterisation of the Rhizoremediation of Petroleum-Contaminated Soil: Effect of Different Influencing Factors

J. C. Tang, R. G. Wang, X. W. Niu, M. Wang, H. R. Chu, and Q. X. Zhou

College of Environmental Science and Engineering, Nankai University/ Key Laboratory of Pollution Processes and Environmental Criteria, Ministry of Education, Tianjin, 300071, China

ABSTRACT

Pilot experiments were conducted to analyse the effect of different environmental factors on the rhizoremediation of petroleum-contaminated soil. Different plant species (cotton, ryegrass, tall fescue and alfalfa), the addition of fertilizer, different concentrations of total petroleum hydrocarbons (TPH) in the soil, bioaugmentation with effective microbial agents (EMA) and plant growth-promoting rhizobacteria (PGPR) and remediation time were tested as influencing factors during the bioremediation process of TPH. The results show that the remediation process can be enhanced by different plant species. The order of effectiveness of the plants was the following: tall fescue > ryegrass > alfalfa > cotton. The degradation rate of TPH increased with increased fertilizer addition, and a moderate urea level of 20 g N (Nitrogen)/m2 was best for both plant growth and TPH remediation. A high TPH content is toxic to plant growth and inhibits the degradation of petroleum hydrocarbons. The results showed that a 5% TPH content gave the best degradation in soil planted with ryegrass. Bioaugmentation with different bacteria and PGPR yielded the following results for TPH degradation: cotton+EMA+PGPR > cotton+EMA > cotton+PGPR > cotton > control. Rapid degradation of TPH was found at the initial period of remediation caused by the activity of microorganisms. A continuous increase of degradation rate was found during the 30–90 days period followed by a slow increase during the 90–150 days period. These results suggest that rhizoremediation can be enhanced with the proper control of different influencing factors that affect both plant growth and microbial activity in the rhizosphere environment

INTRODUCTION

With the development of the economy and petroleum exploration, the contamination of soil with petroleum compounds is a concern worldwide (Banks et al., 2003; Rojo, 2009). Bioremediation of contaminated soil is low cost, causes less interference with the soil structure and has a higher public acceptance than other approaches including soil thermal desorption and soil leaching treatment. There are two different approaches for the bioremediation of petroleum contaminated soil: microbial remediation and phytoremediation. Phytoremediation is a strategy that uses plants to degrade, stabilize, and/or remove soil contaminants. Phytoremediation of hydrocarbons has the potential to be a sustainable waste management technology if it can be proven to be effective in the field (Gurska et al., 2009). Recently, the combination of microbial remediation and phytoremediation has become a general practice in the field treatment of petroleum contaminated soils. This technique can be defined as rhizoremediation, which is a specific type of phytoremediation that involves both plants and their associated rhizosphere microbes. The process can occur naturally or can be actuated by deliberately introducing specific microbes (Gerhardta et al., 2009). The contamination of crude oil results in an immediate change in the bacterial community structure, an increasing abundance of hydrocarbon-degrading microorganisms and a rapid rate of oil degradation, which suggests the presence of a pre-adapted, oil-degrading microbial community and sufficient supply of nutrients (Coulon et al., 2006; Hamamura et al., 2006). The degradation rates of microbial remediation and phytoremediation greatly differ, depending on several conditions. Microbial degradation can be accomplished by different species of microorganisms that are both native to the soil and added as effective degrading strains. The microbial degradation is generally higher than 40% within 1 year of disposal and may be as high as 70% in some cases (Sathishkumar et al., 2008). Influencing factors for microbial remediation include soil moisture content, soil temperature, soil pH, oxygen supply, nutrients, oxidation-reduction potential, soil texture and soil structure (Riser-Roberts, 1998). However, the degradation rate in phytoremediation is generally low; it may be as low as only 9.1%–20% higher than that of the control soil (Brandt et al., 2006; Euliss et al., 2008).

Accordingly, bioaugmentation is needed to introduce effective microorganisms to improve the efficiency of rhizoremediation. The synergistic reaction of the plants and microorganisms, rhizoremediation showed a higher degradation rate of petroleum pollutants than microbial remediation and phytoremediation (Gurska et al., 2009; Xin et al., 2008; Escalante-Espinosa et al., 2005). Several plant species, including ryegrass, sorghum, maize, alfalfa, Bermuda grass, rice, legume, sorghum and beggar ticks, are effective in degrading total petroleum hydrocarbons (TPH) (Nedunuri et al., 2000; Kaimi et al., 2007; Merkl et al., 2005; Shirdam et al., 2008). The TPH content is an important factor of rhizoremediation and influences the bioremediation process. A high TPH content is toxic to both microorganisms and plants. Some plants are sensitive to oil pollution, and plant growth may be greatly reduced in a high-TPH soil (Peng et al., 2009). By using ryegrass and plant growth-promoting rhizobacteria (PGPR) in a rhizoremediation process, the degradation rate in soil with a TPH content of 13% was 61.5% during 3 years of remediation (Gurska et al., 2009). When the TPH content was 5%, the process removed 90% of all TPH fractions from the soil, whereas phytoremediation alone was only able to remove about 50% of the TPH in the same time period (Huang et al., 2005). These results suggest that a high TPH content inhibits plant growth and microbial activity in the rhizosphere environment, which then results in low TPH degradation. Other factors affecting the rhizoremediation process include inoculation, the addition of nutrients, soil organic content, soil depth and salt content (Mishra et al., 2001; Margesin et al., 2003; Lin and Mendelssohn, 1998; Hutchinson et al., 2001; Keller et al., 2008). Despite our understanding of the mechanisms of remediation and the successful results in the laboratory and greenhouse, efforts to translate bioremediation research to the field have been challenging (Gerhardta et al., 2009), which is partially because plant growth in the field is generally different from that under laboratory conditions. Furthermore, the field remediation can be affected by many different factors.

Table 1: Chemical properties of the soil used in the experiment

pH	TPH%	Total N(g/kg)	Total N(g/kg)	Heavy metal (mg/kg)					
				Zn	Cd	Ni	Cu	Pb	Cr
7.9	10	2.75	0.11	666	-	6.5	163	-	12

Currently, systematic research on the influencing factors of rhizoremediation is lacking even though various rhizoremediation techniques for petroleum-contaminated soil have been applied in both laboratory and field experiments. For a better understanding of the mechanisms of remediation and the enhancement of the remediation efficiency, a series of rhizoremediation experiments was conducted and compared to further understand how different factors affect the rhizoremediation process and how the remediation process can be controlled for better disposal of TPH-contaminated soil.

MATERIALS AND METHODS

Petroleum-contaminated soil

Experiments for Different Influencing Factors Of Rhizoremediation

The following five experimental series were designed to study different influencing factors on the effect of *rhizoremediation*. (1) Comparison of different species regarding the remediation of TPH. The following four plant species were selected based on literature reports showing that these plants have been used for the remediation of petroleum pollutants: cotton (*Gossypium hirsutum Linn*), ryegrass (*Lolium perenne L.*), tall fescue (*Festuca arundinacea*) and alfalfa (*Medicago sativa*). The soil TPH content was 5%, and the experiment was carried out in flower pots with 750 g of soil for 150 days. The plants were managed based on their general requirements. The TPH content was tested after the remediation process. (2) Effect of chemical fertilizer addition on the remediation process. Urea was added to the TPH contaminated soil at rates of 0 g N (Nitrogen) /m2 , 5 g N/m2 , 10 g N/m2 , 20 g N/m2 and 30 g N/m2 , using the same soil as described in the previous experiment. Tall fescue was used as the remediation plant, and the TPH content was analysed and plant biomass was weighed after 150 days of remediation. (3) The effect of TPH content on the growth of plants and the remediation effect. TPH contents of 2%, 5% and 10% were prepared, and tall fescue was planted for a period of 150 days. The degradation rates were calculated based on the change of TPH content before and after the remediation. (4) Effect of bioaugmentation on the remediation process. Pot experiment was carried out using 750 g of petroleum-contaminated soil at a TPH concentration of 5%. Cotton was used in this experiment to study the combined effect of plants and EMA, which has not previously been reported. The following five treatments were used: (a) control, (b) planting cotton, (c) planting cotton and the addition of 2% effective microbial agents (EMA), (d) cotton+PGPR, and (e) cotton+EMA+PGPR. The EMA consisted of 2 microbial strains: Acinetobacter radioresistens and Rhodococcus erythropolis. The bacteria were incubated to over 1×10^{10} cfu/g in liquid media and added to the peat in a ratio of 1:4 (w/w) to produce the microbial agents. PGPR contained mainly Azospirillum Brasilence, which was bought from Shanghai Pengxie Co., Ltd. The cotton seeds were soaked in the PGPR solution before planting. TPH content, dehydrogenase activity and PCR-DGGE analyses were carried out on the samples during bioaugmentation process. (5) Degradation of pollutants at different times throughout the rhizoremediation process. The following three treatments were used: (a) control, (b) addition of 2% EMA (Acinetobacter radioresistens and Rhodococcus erythropolis), and (c) addition of 2% EMA with the planting of tall fescue. Samples were taken after days 0, 15, 30, 45, 60, 90 and 150 to study the dynamic change of TPH content during the remediation process. Filter paper was placed at the bottom of the flower pot to cover the drainage hole, and TPH-contaminated soil was added. For cotton planting, 20 seeds were arranged evenly in each pot and covered with 2–3 cm of soil on the top. For tall fescue, ryegrass and alfalfa, 5 g of seed were added evenly to the soil in each pot and covered with 0.5–1 cm of soil on the top. Water was then added to maintain a soil moisture of 60–90% of the maximum water holding capacity. No fertilizer was added during the remediation process. In experiment 2, different amounts of urea were added to the TPHcontaminated soil that was placed in the flower pot, and the soil was then mixed thoroughly.

Analysis of TPH

The TPH content was analysed by air-drying 5-g subsamples of the soil. The samples then underwent ultrasound extraction with 15 mL of chloroform for 15 min followed by centrifugation at 4000 rpm. The supernatant was then filtrated and placed in a flask. The extraction procedure was repeated 3 times, and the extracts were concentrated by drying with a rotary evaporator at 40 °C. The samples were then dried to constant weights at 60 °C, and the flask was reweighed to determine the petroleum hydrocarbon contents.

Dehydrogenase Activity Analysis

The soil was air-dried, and a 5-g sample was placed in a flask. A solution consisting of 5 mL of 0.1% TTC (2,3,5- triphenyltetrazolium Chloride) and 2 mL of 0. 2M Tris-HCl buffer solution (pH 7.6) was added to the flask, which was then mixed and shaken. A blank was prepared with no addition of TTC. The prepared samples were incubated at 37 ∘C for 24 h. After incubation, a 1 M H2SO4 solution was used to stop the reaction, and 5 mL of toluene was added. The flasks were then incubated for 30 min with shaking. After centrifugation, the absorbance of the organic solution was detected at 492 nm, and the dehydrogenase activity was expressed as the amount of TPF (Triphenyl formazan) produced by the reduction of TTC.

Denaturing gradient gel electrophoresis (DGGE) Analysis of Partial 16s RRNA Genes

DNA extraction was carried out using a ZR Soil Microbe DNA Kit™ (Orange, CA). The procedure of the DNA extraction from soil was performed according to the manufacturer's instructions. The 16S rDNA genes were amplified with PCR using the following primers: 357f-GC clamp (forward, 50-CGCCCGCCGCGCGCGGCGGGCGGGGCGGGGGCAC GGGGGGCCTACGGGAGGCAGCAG-30 , Escherichia coli position: 341–357; the underlined sequence corresponds to the GC clamp) and 517r (reverse, 50 - ATTACCGCGGCTGCTGG-30 , Escherichia coli position: 517–534). The reaction solution for the PCR contained 0.5 μl of each primer, 2.5 μl of 10X Ex Taq buffer, 2.5 μl of 2.5 mM dNTP mixture, 2.0 μl of bovine serum albumin (BSA), 1.5 μl of 25 M Mg2+, 0.25 μl of 5 U/ μl Ex Taq DNA polymerase (TaKaRa, Otsu, Japan), 1.0 μl of DNA template and 14.75 μl of ultrapure sterile water. The PCR was carried out by a TaKaRa PCR Thermal Cycler Dice Model TP600 (TaKaRa, Otsu, Japan) with the following conditions for amplification: initial denaturation at 94 ∘C for 3 min, 25 cycles at 1 min each of denaturation at 94 ∘C, 1 min of annealing at 55 ∘C and 2 min of extension at 72 ∘C, followed by a final extension at 72 ∘C for 7 min. DGGE analysis was used to obtain fingerprinting of the PCR products of partial 16S rDNA genes. The denaturing gradient of polyacrylamide ranged from 30% to 60%. The PCR products were used to perform gel electrophoresis in a cell D-code TM system (Bio-Rad laboratories, Hercules, CA, USA) at 60 °C and 200 V for 4 h. The gel was then stained with ethidium bromide (EB) for 30 min and photographed under UV light. The DGGE bands were analysed by Quantity One (Bio-Rad laboratories, Hercules, CA, USA).

RESULTS

Effects of different plant species on the remediation process of TPH

Different plant species have been used in the phytoremediation process of petroleum-contaminated soil. Grass of tall fescue and ryegrass as well as alfalfa are generally used and are effective in enhancing the bioremediation of TPH. Cotton is an economic crop that has adapted to saline and alkaline soils.

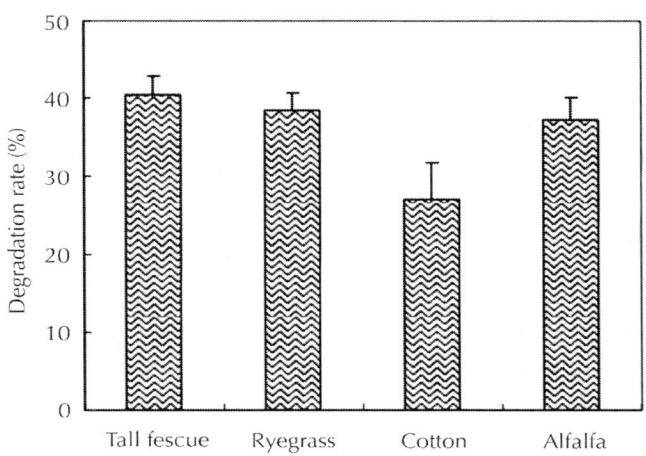

Figure 1: Comparison of the abilities of different plant species tobioremediate TPH

The effects of different plant species on the remediation of TPH are shown in Fig. 1. The degradation of TPH ranged from 33.1% to 48.6%

with the following order of effectiveness of the plants: tall fescue > ryegrass > alfalfa > cotton. The values of degradation in tall fescue and ryegrass were almost the same and were only slightly higher than that of alfalfa. The different degradation rates of TPH in different plants were likely caused by the different physiological functions of roots in the different plants and suggest that the proper selection of plant species is an important strategy in the bioremediation process of TPH.

Effect of chemical fertilizer addition to the remediation process

As both microbial activity and plant growth can be affected by the addition of fertilizer, fertilizer addition is an important factor in determining the efficiency of the bioremediation process. Figure 2a shows the degradation rate of TPH with different addition rates of urea. A positive relationship between the degradation rate of TPH and the addition rate of urea indicates that fertilizer is effective in enhancing the rhizoremediation process of TPH. Figure 2b shows the change of biomass weight with the addition of different amounts of urea. Both wet and dry weights of tall fescue increased with the increase in urea when the application rate was under 20 g N/m^2 . However, when the application rate was increased to 30 g N/m2 , a low biomass weight was found. The highest biomass value of 6.28 g was achieved when the urea addition was 20 g N/m^2 , which was approximately 6 times higher than when the urea addition was 30 g N/m^2

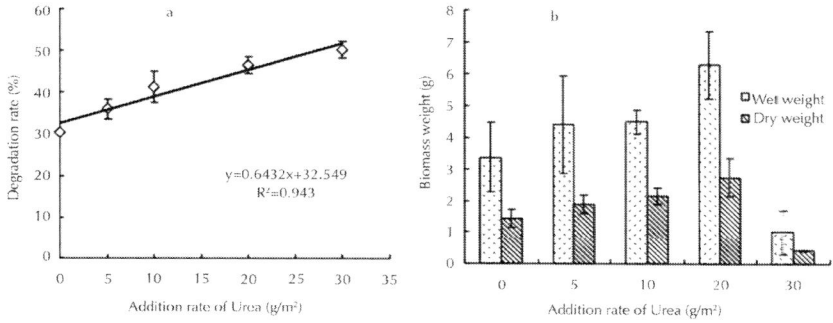

Figure 2: The effect of urea addition on the bioremediation of TPH.

EFFECT OF TPH CONCENTRATION ON PLANT GROWTH ANDREMEDIATION EFFECTIVENESS

Petroleum is toxic to plants, therefore, high concentrations of TPH inhibit plant growth. Figure 3 shows the degradation of TPH at various concentrations. During the phytoremediation process by tall fescue for 150 days, the degradation rates in soil with TPH concentrations of 2%, 5% and 10% was 60.3%, 48.4% and 14.9%, respectively. The degradation rates indicate that low concentrations of TPH are favourable for TPH degradation. However, the degradation amount of THP is the highest when a moderate concentration of THP is present (Fig. 3). Proper evaluation of the TPH concentration should, thus, be considered during the phytoremediation process to achieve the best remediation results.

Effect of Bioaugmentation on the Rhizoremediation Process

The rhizoremediation process is a combined effect of microbial degradation and plant growth. Figure 4 compares the effect of plant growth, addition of EMA and addition of PGPR on the bioremediation process. The TPH degradation order of effectiveness is: cotton+EMA+PGPR > cotton+EMA > cotton+PGPR > cotton > control. The highest degradation rate was 29.8% and was found in the treatment of cotton with the addition of both EMA and PGPR. This treatment showed a 10% higher degradation rate than the control, indicating that bioaugmentation with EMA and PGPR is effective in promoting the rhizoremediation process of TPH. When comparing cotton planting with the control, a 5% higher degradation rate in cotton planting treatment suggested the effectiveness of phytoremediation. However, cotton+EMA and cotton+PGPR planting showed higher degradation rates than those achieved through cotton planting only. Bioaugmentation with petroleum-degrading bacteria is supposed to be able to enhance the rhizoremediation process. Dehydrogenase activity was higher in the four treatments than in the control. The highest value of dehydrogenase activity was achieved with the addition of EMA. However, PGPR did not improve the dehydrogenase activity during the remediation process. Based on the DGGE analysis results shown in Fig. 5a, more complex microbial communities were found in lanes 2–6 compared to that in lane 1.

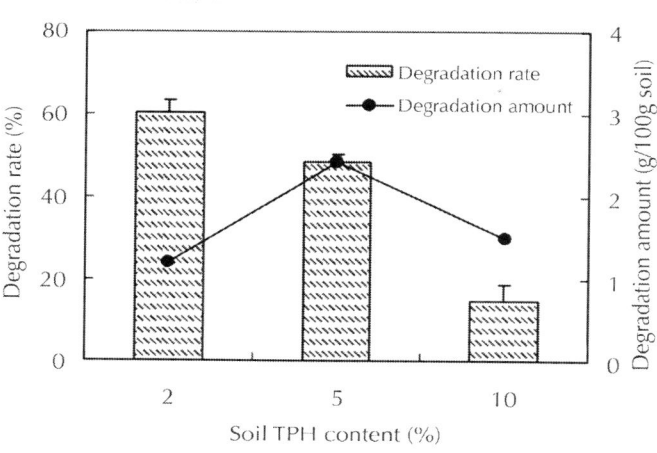

Figure 3: The effect of soil TPH concentration on the degradation rate.

These results indicate that the rhizoremediation of TPH with cotton and different microorganisms allows the proliferation of the complex microbial community. Some new bands developed during the remediation process. Common bands C1-C13 can be found in all samples of lanes 1–6. Special bands S1-S8 are only observed in certain lanes. S1 was found in lanes 2–6 but not in lane1, S2 was only found in lane 1, S3 was only found in lanes 4 and 5, and S5 was only found in lane 3. The cluster analysis shown in Fig. 5b indicates that the microbial community of the control group was different from that of the cotton planting group. However, the two treatments with the addition of PGPR were grouped into one category by the cluster analysis.

Degradation of TPH Pollutants at Different Time Throughout the Rhizoremediation Process

The rhizoremediation process with tall fescue can be divided into three periods. A rapid increase in the degradation rate was found at the initial period of remediation of days 0–30, a continuous increase in the degradation rate was found in the following period of days 30–90, and a slight increase in the degradation rate was found from days 90 to 150. At days 15, the relative degradation rates were: EMA > EMA+plant > control. The degradation rate gradually increased in the EMA+plant treatment and became higher than that of EMA in TPH degradation after day 30. An enhanced remediation effect by planting tall fescue coupled with the addition of EMA (7% higher than the control and EMA sample) was found after remediation days 90 (Fig. 6).

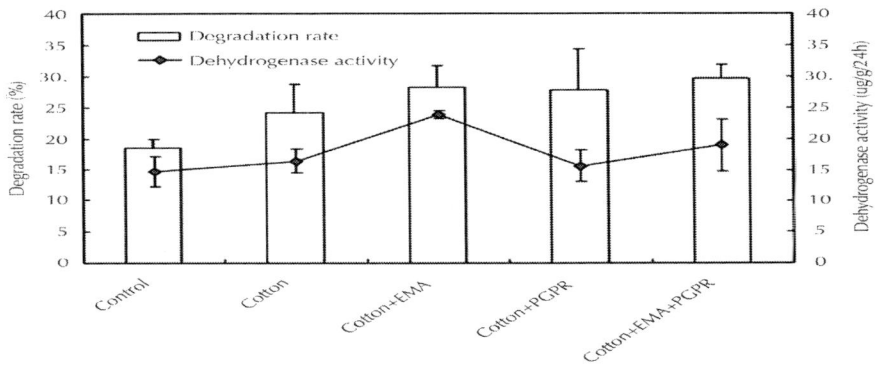

Figure 4: The effect of EMA and PGPR bacteria on the remediation process.

DISCUSSION

Regulating the factors influencing the rhizoremediation process is important for achieving an improved degradation of TPH. In this study, plant species, fertilizer addition, TPH concentration, inoculation of microorganisms and remediation time were considered to be the main factors influencing rhizoremediation and were studied in detail. In our research, tall fescue, ryegrass, cotton and alfalfa were compared in rhizoremediation process. These plants were chosen because they grow in a wide range of soil conditions and can withstand high saline-alkali stress, which is common in oil fields in many parts of China. Hydrocarbon contamination is known to significantly reduce the growth of the plants, and the corresponding TPH degradation efficiency of the different plant species is reported to differ widely (Shircam et al., 2008; Euliss et al., 2008). The TPH degradation rate among various plant species depends on the microbial population in the rhizosphere of these plants. Many different plant species have been reported to be effective in the remediation of TPH-contaminated soil, including grass, alfalfa, poplar, and ryegrass (Phillips et al., 2009; Euliss et al., 2008; Kechavarzi et al., 2007). It was reported that perennial ryegrass and alfalfa increased the number of rhizosphere bacteria in hydrocarbon-contaminated soil (Kirk et al., 2005). Tall fescue also showed a high TPH degradation rate in petroleum-contaminated soil (Ezzatian et al., 2009). Cotton can grow under conditions of approximately 0.5% salinity (Ashraf and Ahmad, 2000; Sacchi et al., 2000) and can be grown in the saline-alkaline oil fields in China. However, the degradation rate of TPH by cotton is lower than that of the other three plant species in this study. It was reported by Hou et al. (2001) that rooting intensity (mg root kg^{-1} soil) is the key factor leading to high TPH loss rates, and root development is crucial in evaluating the phytoremediation potential. The well-developed root system of grass is advantageous in enhancing petroleum degradation.

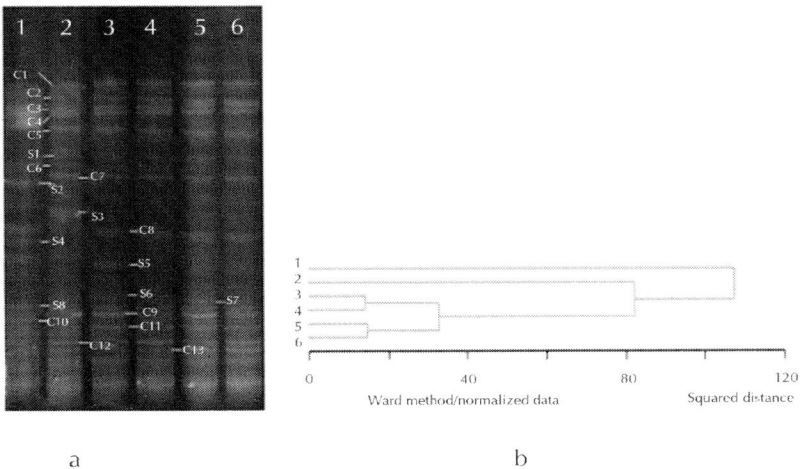

a   b

Figure 5: DGGE analysis results from the TPH bioremediation process (a) and cluster analysis (b). Soil before remediation (1), control (2), cotton (3), cotton+EMA (4), cotton+PGPR (5), and cotton+PGPR+EMA (6).

In contrast, alfalfa has an advantage because of its ability to recruit nitrogen fixing microbes in the soil (Kelner et al., 1997). In addition, cotton is important in phytoremediation because it produces products with high economic value, which might help to sustain the long-term application of field phytoremediation (Banuelos, 2006). It is important to rank the different plant species in TPH degradation. However, the degradation rate is not the only factor used in selecting phytoremediation plants. Plant species should be selected based on the soil condition and remediation purpose. Microbial degradation rates in TPH-contaminated soil were more affected by soil properties and the chemical characteristics of the contaminant than the presence of roots (Song et al., 2004). This result indicates that research should be focused more on the effects of different soil properties on TPH remediation rather than on the selection of different plant species. The addition of NPK fertilizer and compost was shown to greatly enhance hydrocarbon degradation (Palmroth et al., 2006). Furthermore, a positive correlation between TPH degradation rate and fertilizer addition supports this conclusion and suggests that a moderate amount of fertilizer is required for both better plant growth and higher TPH degradation. The inhibition of plant growth at a urea content of 30 g N /m^2 is likely caused by salt toxicity. However, the microbial degradation of TPH was not affected at this urea concentration, as bacteria are tolerant to higher salt contents. Although phosphorus (P) is another factor that might affect plant and microbial growth, our results suggest that N is the most crucial factor in the rhizoremediation process. The effect of TPH concentration on the degradation of TPH is likely caused by the toxicity of hydrocarbons on the plants and rhizosphere microorganisms. TPH concentration was the major determinant of total bacterial abundance and affected the abundance of hydrocarbon degraders (Nie et al. 2009),

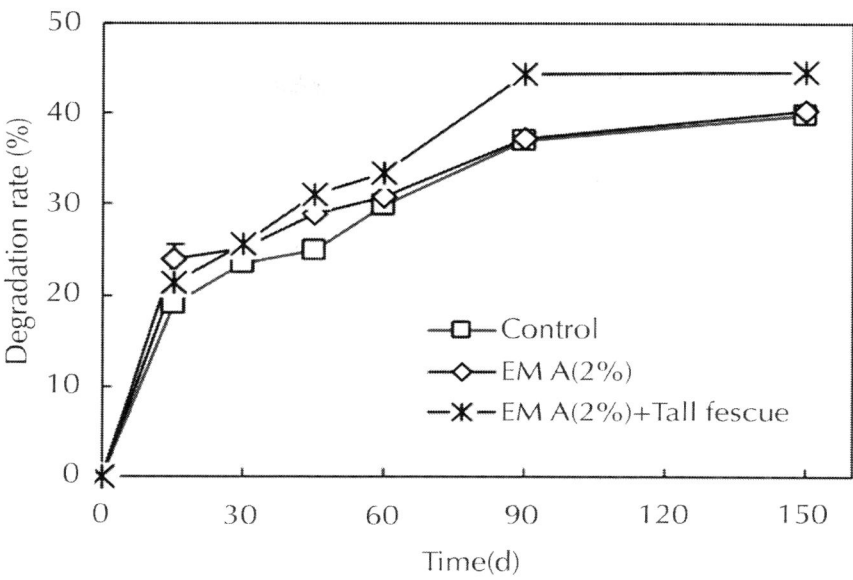

Figure 6: Degradation of TPH pollutants at different time points during the rhizoremediation process with tall fescue.

The bioremediation of hydrocarbons could be carried out successfully at TPH concentrations of 10–13% (Del Panno et al., 2005; Gurska et al., 2009), although there is no report of the maximum permissible TPH content for bioremediation. In addition, the optimum TPH content for rhizoremediation is still a question of concern. Earthworms were 1.4 to 14 times more sensitive to TPH than microorganisms and 1.3 to >77 times more sensitive to TPH than plants (Dorn et al., 1998), meaning that plants can withstand higher concentrations of TPH than other organisms. A preferential degradation of saturated hydrocarbons was found by Peng et al. (2009), during a phytoremediation trial of TPH using *Mirabilis Jalapa* L. Light oil fractions were generally more toxic than heavy oil fractions (Dorn et al., 1998), and the preferential degradation of light oil generally causes a decrease in soil toxicity during the TPH degradation process. On the other hand, the degradation of older soil contaminated by TPH is generally more difficult than freshly contaminated soil. Research shows greater rates of hydrocarbon loss in soils containing fresh petroleum products compared with the aged products (Parker and Burgos, 2001).

Our results indicate that a moderate TPH concentration (about 5%) is favourable for hydrocarbon degradation, which means that biological remediation is most effective in a moderate TPH concentration of contaminated soil immediately after a spill. However, rhizomediation can be carried out successfully even under high TPH concentrations, indicating that further research associated with reducing the toxicity of TPH and enhancing the rhizoremediation process for an improved remediation is necessary. As TPH reduction is positively correlated with culturable hydrocarbon degraders (Phillips et al., 2006), it is possible that the addition of effective microorganisms enhances the rhizoremediation process of TPH (Gurska et al., 2009; Mishra et al., 2001). Different plant species may introduce Biogeosciences, 7, 3961–3969, 2010 www.biogeosciences.net/7/3961/2010/J. C. Tang et al.: Rhizoremediation of petroleum contaminated soil 3967 different microbial communities and increase the total culturable microbial amount during their growth processes. Cotton roots can be colonised by a variety of microorganisms including *Fusarium culmorum*, F. solani, F. *oxysporum, Macrophomina phaseoli* and Bacillus sp. (Ghaffar and Parveen, 1969). Some bacteria, such as the *Rhodococcus* sp. strain, can grow at the oil-water interface and produce a mycolic acid-containing capsule to enhance TPH degradation (Van Hamme and Ward, 2001). A high degradation rate of 85.67% within 120 days was reached by combining cotton with native microorganisms. The degradation rates were higher when cotton was combined with native microorganisms than when other plant species, such as sunflowers, bermuda grass and sudan grass, were combined with native microorganisms (Liu et al., 2009). The results of the DGGE analysis conducted here show that cotton growth increases microbial diversity and alters the microbial community structure. The inoculated microorganisms will remain active during plant growth conditions and colonise in the root systems of plants. In some cases, the population of introduced strains may remain stable even after one year (Mishra et al., 2001). However, researchers reported that the introduction of exotic microorganisms did not improve the remediation and that the inoculation of oil-contaminated sites with nonindigenous species is likely to fail (Li et al., 2002; *Thomassin-Lacroix* et al., 2002; Cavalca et al., 2002). It is suggested that the effect and efficiency of inoculated microorganism depend on the environmental conditions such as the existence of indigenous microorganisms, nutrient level and amount of inoculation. Microbial degradation combined with phytoremediation will further improve the degradation efficiency. As microbial degradation occurred at the early stage of remediation, an early period of microbial remediation followed by phytoremediation is likely to be effective in the field remediation practice. It generally takes several months for the bioremediation of TPH to reach a reasonable end point. Our results suggest that the degradation rate of TPH is different at different time during the bioremediation process. It is possible that native microorganisms may also develop at the initial period of bioremediation as indicated by the high degradation of TPH in the control in experiment 5. While several factors influence rhizoremediation, TPH content is the most important factor. Other factors that can affect the degradation rate include the amount of microbial agents and the addition of fertilizer. As microbial degradation is most effective at the initial period of remediation, a second addition of the microbial agents may be a good strategy during the course of remediation. The regulation of soil conditions such as pH, moisture content and nutrient content is the most important measure for the growth of native microorganisms and plants during the initial phase of rhizoremediation. It is also recommended to use plants with fast growth rates and plants that can become strong enough, such that after only one month they can take over the role of the inoculated microorganisms in maintaining the TPH degradation activity for the long term. Thus, the selection of fast growing plants that have long growth periods is an important factor to consider in the combination of microbial remediation and phytoremediation. After 150 days, the degradation rate is stable even when microbial remediation is combined with the planting of tall fescue. Further research should be performed to enhance the bioremediation by multi-process phytoremediation systems such as the addition of fertilizer, management of plant growth and increasing aeration of the rhizosphere environment.

CONCLUSIONS

Grass plants like tall fescue and ryegrass are better for the bioremediation of TPH than alfalfa and cotton, as these plants showed a higher TPH degradation rate. The addition of urea enhanced the degradation of TPH, and a positive correlation between the amount of urea added and the degradation rate of TPH was found. A high

TPH content inhibited the bioremediation process likely because of the toxicity of TPH to the plant and bacteria. The optimum TPH concentration for rhizoremediation was determined to be approximately 5%. Bioaugmentation during the phytoremediation process with different bacteria including PGPR was shown to enhance the rhizoremediation process with a degradation rate of 29.8% after 150 days. Bioaugmentation was effective during the first 30 days, and plant growth enhanced the TPH degradation after several months of remediation. The results suggest that rhizoremediation can be enhanced by controlling the factors that affect both plant growth and microbial activity in the rhizosphere environment. Because a long time period is needed to conduct bioremediation processes, it is difficult to control all the influencing factors. The TPH content is crucial for rhizoremediation because of its toxicity to plants. Plant selection is also important, as crop management will greatly differ for different plant species. Finally, fertilizer and EMA addition should be applied in the remediation field for better degradation of TPH.

REFERENCES

1.   Ashraf, M. and Ahmad, S.: Influence of sodium chloride on ionaccumulation, yield components and fibre characteristics in salttolerantand salt-sensitive lines of cotton (Gossypium hirsutumL.), Field Crop. Res., 66(2), 115–127, 2000.

2.   Banks, M. K., Mallede, H. and Rathbone, K.: Rhizosphere microbial characterization in petroleum-contaminated soil, Soil Sediment Contam., 12, 371–385, 2003.

3.   Banuelos, G. S.: Phyto-products may be essential for sustainability and implementation of phytoremediation, Environ. Pollut., 144(1), 19–23, 2006.

4.   Brandt, R., Merkl, N., Schultze-Kraft, R., Infante, C., and Broll, G.: Potential of vetiver (Vetiveria zizanioides (L.) Nash) for phytoremediation of petroleum hydrocarbon-contaminated soils in Venezuela, Int. J. Phytoremediat., 8, 273–284, 2006.

5.   Cavalca, L., Colombo, M., Larcher, S., Gigliotti, C., Collina, E., and Andreoni, V.: Survival and naphthalene-degrading activity of Rhodococcus sp. strain 1BN in soil microcosms, J. Appl. Microbiol., 92, 1058–1065, 2002.

6. Coulon, F., McKew, B. A., Osborn, A. M., McGenity, T. J., and Timmis, K. N.: Effects of temperature and biostimulation on oildegrading microbial communities in temperate estuarine waters, Environ. Microbiol., 9, 177–186, 2006.

7. Del Panno, M. T., Morelli, I. S., Engelen, B., and Berthe-Corti, L.: Effect of petrochemical sludge concentrations on microbial communities during soil bioremediation, FEMS Microbiol. Ecol., 53(2,1), 305–316, 2005.

8. Dorn, P. B., Salanitro, J. P., Wisniewski, H. L., and Vipond, T. E.: Assessment of the acute toxicity of crude oils in soils using earthworms, microtox®, and plants, Chemosphere, 37, 845–860, 1998.

9. Escalante-Espinosa, E., Gallegos-Martinez, M. E., Favela-Torres, E., and Gutierrez-Rojas, M.: Improvement of the hydrocarbon phytoremediation rate by Cyperus laxus Lam. inoculated with a

10. microbial consortium in a model system, Chemosphere, 59, 405–413, 2005.

11. Euliss, K., Ho, C. H., Schwab, A. P., Rock, S., and Banks, A. K.: Greenhouse and field assessment of phytoremediation for petroleum contaminants in a riparian zone, Bioresource Technol., 99, 1961–1971, 2008.

12. Ezzatian, R., Voussoughi, M., Yaghmaei, S., Abedi-Koupai, J., Borghei, M., and Ghafoori, S.: Effects of Puccinellia Distansn and Tall Fescue on Modification of C/N Ratios and Microbial Activities in Crude Oil-Contaminated Soils, Petrol Sci. Technol. 27, 452–463, 2009.

13. Gerhardta, K. E., Huang, X. D., Glicka, B. R., and Greenberg, B.M.: Phytoremediation and rhizoremediation of organic soil contaminants:Potential and challenges, Plant Sci. 176, 20–30, 2009.

14. Gurska, J., Wang, W. X., Gerhardt, K. E., Khalid, A. M., Isherwood, D. M., Huang, X. D., Glick, B. R., and Greenberg, B. M.:Three year field test of a plant growth promoting rhizobacteria enhanced phytoremediation system at a land farm for treatment of hydrocarbon waste, Environ. Sci. Technol., 43(12), 4472–4479, 2009.

15. Ghaffar, A. and Parveen, G.: Colonization of cotton roots by soil micro-organisms, Mycopathologia, 38, 373–376, 1969.

16. Hamamura, N., Olson, S. H., Ward, D. M., and Inskeep, W.P.: Microbial Population Dynamics Associated with Crude-Oil Biodegradation in Diverse Soils, Appl. Environ. Microbiol., 72, 6316–6324, 2006.

17. Hou, F. S., Milke, M. W., Leung, D. W., and MacPherson, D.J.: Variations in phytoremediation performance with dieselcontaminated soil, Environ. Technol., 22(2), 215–22, 2001.

18. Huang, X. D., El-Alawi, Y., Gurska, J., Glick, B. R., and Greenberg, B. M.: A multi-process phytoremediation system for decontamination of persistent total petroleum hydrocarbons (TPHs) from soils, Microchem. J., 81, 139–147, 2005.

19. Hutchinson, S. L., Banks, M. K., and Schwab, A. P.: Phytoremediation of aged petroleum sludge: Effect of inorganic fertilizer, J. Environl. Qual., 30, 395–403, 2001.

20. Kaimi, E., Mukaidani, T., and Tamaki, M.: Screening of twelve plant species for phytoremediation of petroleum hydrocarboncontaminated soil, Plant Product Sci., 10, 211–218, 2007.

21. Kechavarzi, C., Pettersson, K., Leeds-Harrison, P., Ritchie, L., and Ledin, S.: Root establishment of perennial ryegrass (L-perenne) in diesel contaminated subsurface soil layers, Environ. Pollut., 145, 68–74, 2007.

22. Keller, J., Banks, M. K., and Schwab, A. P.: Effect of soil depth on phytoremediation efficiency for petroleum contaminants, J.Environ. Sci. Heal. A, 43, 1–9, 2008.

23. Kelner, D. J., Vessey, J. K., and Entz, M. H.: The nitrogen dynamics of 1-, 2- and 3-year stands of alfalfa in a cropping system, Agr. Ecosyst. Environ., 64(1), 1–10, 1997.

24. Kirk, J. L., Klironomos, J. N., Lee, H., and Trevors, J. T.: The effects of perennial ryegrass and alfalfa on microbial abundance and diversity in petroleum contaminated soil, Environ. Pollut., 133, 455–465, 2005.

25. Li, P., Sun, T., Stagnitti, F., Zhang, C., Zhang, H., Xiong, X., Allinson, G., Ma, X., and Allinson, M.: Field-Scale Bioremediation of Soil

Contaminated with Crude Oil, Environ. Eng. Sci.,19, 277–289, 2002.

26. Lin, Q. X. and Mendelssohn, I. A.: The combined effects of phytoremediation and biostimulation in enhancing habitat restoration and oil degradation of petroleum contaminated wetlands,Ecol. Eng., 10, 263–274, 1998.

27. Liu, J. C., Cui, Y. S., Zhang, Y. P., and Zou, S. Z.: Effect of Plants and Microorganisms on Remediation of Petroleum Contaminated Soil, J. Ecol. Rural Environ., 25, 80–83, 2009.

28. Margesin, R., Labbe, D., Schinner, F., Greer, C. W., and Whyte, L. 'G.: Characterization of Hydrocarbon-Degrading Microbial Populations in Contaminated and Pristine Alpine Soils, Appl. Environ. Microbiol., 69, 3085–3092, 2003.

29. Merkl, N., Schultze-Kraft, R., and Infante, C.: Assessment of tropical grasses and legumes for phytoremediation of petroleumcontaminated soils, Water Air Soil Pollut., 165, 195–209, 2005.

30. Mishra, S., Jyot, J., Kuhad, R. C., and Lal, B.: Evaluation of inoculum addition to stimulate in situ bioremediation of oily-sludgecontaminated soil, Appl. Environ. Microbiol., 67, 1675–1681,

31. 2001.

32. Nedunuri, K. V., Govindaraju, R. S., Banks, M. K., Schwab, A. P., and Chens, Z.: Evaluation of phytoremediation for field-scale degradation of total petroleum hydrocarbons, J. Environ. Eng.-ASCE, 126, 483–490, 2000.

33. Nie, M., Zhang, X. D., Wang, J. Q., Jiang, L. F., Yang, J., Quan, Z. X., Cui, X. H., Fang, C. M., and Li, B.: Rhizosphere effects on soil bacterial abundance and diversity in the Yellow River

34. Deltaic ecosystem as influenced by petroleum contamination and soil salinization, Soil Biol. Biochem. 41(12), 2535–2542, 2009.

35. Palmroth, M., Koskinen, P. E. P., Pichtel, J., Vaajasaari, K., Joutti, A., Tuhkanen, T. A., and Jaakko Puhakka, A.: Field-Scale Assessment of Phytotreatment of Soil Contaminated with Weathered Hydrocarbons and Heavy Metals, J. Soils Sediments, 6(3), 128–136, 2006.

36. Parker, E. F. and Burgos, W. D.: Degradation Patterns of Fresh and Aged Petroleum-Contaminated Soils, Environ. Eng. Sci., 16, 21–29, 1999.

37. Peng, S., Zhou, Q., Cai, Z., and Zhang, Z.: Phytoremediation of petroleum contaminated soils by Mirabilis Jalapa L. in a greenhouse plot experiment, J. Hazard. Mater., 168, 1490–1496, 2009.

38. Phillips, L. A., Greer, C. W., Farrell, R. E., and Germida, J. J.:Field-scale assessment of weathered hydrocarbon degradation by mixed and single plant treatments, Appl. Soil Ecol., 42, 9–17, 2009.

39. Phillips, L. A., Greer, C. W., and Germida, J. J.: Culture-based and culture-independent assessment of the impact of mixed and single plant treatments on rhizosphere microbial communities in hydrocarbon contaminated flare-pit soil, Soil Biol. Biochem., 38, 2823–2833, 2006.

40. Riser-Roberts, E.: Remediation of Petroleum Contaminated Soils: Biological, Physical, and Chemical Processes, St. Lucie Press, Boca Raton, FL, 1998.

41. Rojo, F.: Degradation of alkanes by bacteria, Environ. Microbiol. 11, 2477–2490, 2009.

42. Sacchi, G. A., Abruzzese, A., Lucchini, G., Fiorani, F., and Cocucci, S.: Efflux and active re-absorption of glucose in roots of cotton plants grown under saline conditions, Plant Soil, 220, 1–11, 2000.

43. Sathishkumar, M., Binupriya, A. R., Baik, S. H., and Yun, S. E.: Biodegradation of crude oil by individual bacterial strains and a mixed bacterial consortium isolated from hydrocarbon contaminated areas, Clean-Soil Air Water, 36, 92–96, 2008.

44. Shirdam, R., Zand, A. D., Bidhendi, G. N., and Mehrdadi, N.: Phytoremediation of hydrocarbon-contaminated soils with emphasis on the effect of petroleum hydrocarbons on the growth of plant species, Phytoprotection, 89, 21–29, 2008.

45. Song, Y. F., Song, X. Y., Zhang, W., Zhou, Q. X., and Sun, T. H.: Issues concerned with the bioremediation of contaminated soils, Huan Jing Ke Xue, 25, 129–133, 2004.

46. Thomassin-Lacroix, E., Eriksson, M., Reimer, K., and Mohn, W.: Biostimulation and bioaugmentation for on-site treatment of weathered diesel fuel in Arctic soil, Appl. Microbiol. Biotechnol., 59, 551–556, 2002.

47. Van Hamme, J. D. and Ward, O. P.: Physical and Metabolic Interactions of Pseudomonas sp. Strain JA5-B45 and Rhodococcus sp. Strain F9-D79 during Growth on Crude Oil and Effect of a

48. Chemical Surfactant on Them, Appl. Environ. Microbiol., 67, 4874–4879, 2001.

49. Xin, L., Li, X. J., Li, P. J., Li, F., Lei, Z., and Zhou, Q. X.: Evaluation of plant-microorganism synergy for the remediation of diesel fuel contaminated soil, Bull. Environ. Contam. Toxicol., 81, 19–24, 2008.

Chapter 10

Comparative Study on Sulphur Reduction From Heavy Petroleum - Solvent Extraction and Microwave Irradiation Approach

Abdullahi Dyadya Mohammed[1], Abubakar Garba Isah[1],
Musa Umaru[1], Shehu Ahmed[1], Yababa Nma Abdullahi[2]

[1]Department of Chemical Engineering, Federal University of Technology, P.M.B 65, Minna, Nigeria.

[2]National Petroleum Investment Management Services (Nigeria National Petroleum Corporation), Lagos, Nigeria.

ABSTRACT

Sulphur- containing compounds in heavy crude oils are undesirable in refining process as they affect the quality of the final product, cause catalyst poisoning and deactivation in catalytic converters as well as causing corrosion problems in oil pipelines, pumps and refining equipment aside environmental pollution from their combustion and high processing cost. Sulphur reduction has being studied using microwave irradiation set at 300W for 10 and 15minutes and oxidative-solvent extraction method using n- heptane and methanol by 1:1, 1:2 and 1:3 crude- solvent ratios after being oxidized with hydrogen peroxide, H_2O_2 oxidants. Percentage sulphur removal with n- heptane solvent by 1:1 and 1:2 are 81.73 and 85.47%; but extraction using methanol by different observed ratios gave less sulphur reduction. Indeed when microwave irradiated at 300W for 10 and 15minutes, 53.68 and 78.45% reduction were achieved. This indicates that microwave irradiation had caused oxidation by air in the oven cavity and results to formation of alkyl radicals and sulphoxide from sulphur compound in the petroleum. The prevailing sulphur found in the crude going by FT-IR results is sulphides which oxidized to sulphoxide or sulphones. It is clear that sulphur extraction with heptane is more efficient than microwave irradiation but economically due to demands for solvent and its industrial usage microwave irradiation can serve as alternative substitute for sulphur reduction in petroleum. Sulphur reduction by microwave radiation should be up- scaled from laboratory to a pilot plant without involving extraction column in the refining.

INTRODUCTION

Heavy petroleum having been dense with high viscosity contains many varieties of complex hydrocarbon molecules along with organic impurities containing sulphur, nitrogen, polycyclic aromatics and asphaltenes, heavy metals and metal salts [1]. However, most of heavy petroleum processed in our refineries is often foreign crude imported from either Middle East or Russian country. It has been known that heavy crude impurities are in concentrated form. Sulphur has been recognized as one of the major impurity which affects the quality of the final refined product and often contributes to high processing cost. Parallel to this, sulphur containing compounds in the crude oil creates many operational issue and also pollutes air when products burn. Sulphur in petroleum poisons the catalyst and causing catalyst deactivation [1, 2].

According to Erikh et al. [3], higher sulphur products leads to increased emissions of sulphur dioxides and resultant acid rain thus giving poor performance of vehicle emissions control systems. Refining processes encountered corrosion problems when processing crude oil with appreciable amount of sulphur and sulphur- containing compounds. Sulphur compounds are highly objectionable in commercial products on account of their unpleasant smell or bad odour [4]. Each crude oil has different amounts and types of sulphur compounds, but as a rule the proportion, stability, and complexity of the compounds are greater in heavy crude oil fractions. The combustion of petroleum products containing sulphur compounds produces undesirables such as sulphuric acid and sulphur dioxide [5]. Conventionally, most refineries adopted hydrotreating processes or hydrodesulphurization (HDS) reaction which involves catalytic conversion of various sulphur compounds to nonsulphur containing materials in the presence of hydrogen. The production of low levels of sulphur- containing compounds product therefore required the application of severe operating conditions, the use of active catalyst [6] and possible innovative technology.

Microwave irradiation is attracting researcher's attention as a tool for facilitating chemical reactions. Indeed, microwave energy, with a frequency of 2.45GHz, is well known to have a significant effect on the rates of a variety of processes. The number of reported applications, especially in the food industry, is increasing rapidly [7, 8]. Microwaves interact strongly with some (polar) materials and weakly with others. Energy absorption varies depending upon microwave operation frequency, sample composition, dielectric properties of the materials and temperature of operation. Dielectric properties of a material or combination of materials affect how the transmitted microwaves react with the materials in the system.

The dielectric properties of crude oil vary according to the type and composition since crude oil has a low dielectric constant (2 to 3) and thus weakly absorbs microwaves. The dielectric factor and the loss factor are not constant because they vary with temperature, frequency (power), and moisture content [9]. Several literatures reported that microwaves heating has effectively shortened the reaction time and dramatically probe optimum production conversion to about 84-86.3% [2, 6, 10].

Microwave heating has an incontestable place in analytical and organic laboratory practices as a very effective and non-polluting method of activation and heating. Two major effects are responsible for the heating, both resulting from interaction of electric field component of the microwaves with the sample [11, 12]. Dipole interactions occur with polar molecules. The polar ends of a molecule tend to align themselves and oscillate in step with the oscillating electrical field of the microwaves. Collisions and friction between the moving molecules result in heating. However, the more polar a molecule, the more effectively it will couple with (and be influenced by) the microwave field. Ionic conduction is only minimally different from dipole interactions. Ions can couple with the oscillating electrical field of the microwaves [13, 14]. The effectiveness or rate of microwave heating of an ionic solution is a function of the concentration of ions in solution [15]. The application of microwave heating to desulphurization reactions has precedent in the patent literature [6].

SULPHUR- CONTAINING COMPOUND

Sulphur compounds are highly polar molecules with sulphur at the negative end, and hydrogen and carbon at the positive end. Zuloaga *et al.*, [16] reported that the more polar a solvent, the quicker its temperature shoots up. According to Erikh *et al.* [3] about 70 to 90% of sulphur containing compounds is concentrated in the fuel oil. Mercaptans and disulphides are chiefly contained in the gasoline and kerosene fraction whereas cyclic sulphides and polycyclic sulphur compounds are contained in the heavier molecular weight hydrocarbons. These get into the kerosene and oil fractions in the distillation of petroleum. The sulphides and mercaptans from C2 to C7 have low boiling points (37-150°C) and easily oxidized or decomposed on heating above the specified temperature range to form sulphides or sulphoxides and with the evolution of H_2S [3].

In addition to this, even air or weak oxidizing agents oxidized sulphides and mercaptans to disulphides. Many researchers observed that the most probable types of sulphur- containing compounds like polycyclic sulphur compounds are more complex to isolate completely but more often than not can be reduced to an acceptable minimum. The aim of the study is to find out the possibility of using microwave irradiation to reduce sulphur compounds in imported foreign crude petroleum; hence sulphur- containing compounds can thermally decomposed (or oxidized) on exposure to a low temperature conditions.

METHODOLOGY

Equipment and materials

- Sulphur- in Oil Analyser (model number XLSA-20), Thermo Nicolet IR Spectrometer (100 FT-IR) System, Magnetic heater and Stirrer with temperature regulator, Thermometer and Sample bottles, Pyrex 25ml Measuring Cylinder and 250ml separating funnel and Domestic Microwave oven (Model: HH7206R).

- Urals heavy crude oil (Russian), Methanol (BDH- Chemical, Poole, England), n- Heptane (Kemie Labs, India), Ethanoic acid (BDH- Chemical, Poole, England) and Aluminium oxide, Al_2O_3 and Hydrogen Peroxide (May & Baker, Dagenham, England).

Procedure for Oxidation of Crude Oil

50ml of crude oil sample was poured into a 250ml conical flask with 5ml of organic acid (ethanoic acid) and a magnetic bar was placed inside for proper stirring. The mixture was heated at 90°C, the 5ml solution of H_2O_2 aqueous was added in drop wise and with constant stirring over 15minutes to oxidize sulphur compounds in the sample.

Procedure for Sulphur Compounds Extraction

The oxidized crude oil sample was mixed in a 250ml separating funnel with each solvent (methanol and n- heptane) at room temperature and atmospheric pressure to obtained biphasic mixture in the ratio 1:1

(20ml of crude oil sample to 20ml of solvent), 1:2 (20ml of crude oil sample to 40ml of solvent) and 1:3 (20ml of crude oil sample to 60ml of solvent). Thereafter, the mixtures were left to achieve extraction equilibrium for 30minutes to form two- phase separation. The crude oil phase was decanted carefully from the solvent- sulphur phase and finally analyzed for total sulphur content.

Procedure for Microwave Desulphurization

50ml of crude oil sample was poured into a 250ml pyrex glass with aluminium oxide, Al_2O_3 (microwave absorber) and placed in the domestic microwave oven (Model: HH7206R). The microwave power was adjusted to 300W. The timer was turned on first to its maximum (60minutes) then back to the desired microwave irradiation time 10minutes. The operation began as soon as the oven door was firmly closed to activate the sample energy. The procedure was repeated by setting microwave irradiation time to 15minutes. After the set times elapsed, microwave irradiated crude oil samples were analyzed respectively.

Procedure for Sulphur- in- Oil Analyzer

The sample was poured into a disposable sample container to three quarter of its capacity. This is to ensure the X- ray passes through the test sample in order to give accurate sulphur counts. The sample was then covered with X- ray transparent plastic film window. The power was switched on and light up the X- ray lamp within seconds. The analyzer gives three different readings at 30seconds intervals. The readings were recorded and average sulphur content was determined in percentage sulphur by weight.

Procedures for Thermo Nicolet IR Spectrometer

A drop of desulphurized crude oil sample obtained was placed between two plates of potassium bromide, KBr salt (transparent to infrared light) and squeeze to remove any trapped air in order to form a thin film (sand witched) between the plates otherwise light cannot pass through. The plates were placed in the sample holder and positioned in the standard sample compartment of the spectrometer and then FT-IR spectra were obtained under 2minutes data collection period with a spectral range from 4000.00 to 406.75cm-1 at a resolution of 4cm-2 [2, 17].

RESULT AND DISCUSSIONS

Table 1 shows the average sulphur content in the crude oil before and after extraction using n- heptane solvent by 1:1, 1:2 and 1:3 ratios as well as under microwave irradiation. Sulphur is the other parameter focused for determining the price of crude oil and refining processing cost. The total sulfur content is expressed as a percentage of sulfur by weight, and it varies from less than 0.1% to greater than 5% depending on the type and source of hydrocarbons respectively [18]. The total sulphur content in the untreated crude oil was 1.494wt%. After oxidation with oxidants, H_2O_2, it reduces by 25.23% to 1.117wt%. This is evidence that sulphur- containing compounds in the crude oil oxidized to sulphoxide or corresponding sulphone.

However, sulphur extraction with heptane by 1:2 gave high percentage reduction (85.47%) to 0.217wt%; this was closely followed by 1:1 ratio where 81.73% reduction to 0.273wt% was achieved. As observed, approximately the same 79.0% (to the nearest whole number) sulphur reduction were achieved by being irradiated at 300W for 15minutes and by using n- heptane to crude oil 1:3 ratio. This inferred that microwave has the potential to reduce sulphur in the crude oil. Most of sulphur- containing compounds found in the crude oil such as thiol (mercaptans), thiophenes, benzothiophenes, dibenzothiophenes, sulphoxide or sulphone among others are polar compounds. Sudipa *et al.* [19] observed that, much greater polarity of the sulphoxide groups moves from oxidized component to the asphaltene.

Indeed, according to Zuloaga *et al.* [16], microwave can turned to heat if they interact with a polar material and causes it to rotate and the quicker its temperature shoot up. There is every possibility some low-boiling sulphur compounds in the crude when heated can oxidized by air in the oven cavity to form sulphoxide and corresponding sulphur oxide, SO_2 as well as hydrogen sulphide, H_2S due to radicals formation and recombination [20]. It is clear that heptane extraction by 1:1 and

1:2 ratios are efficient in sulphur extraction than microwave irradiation treatment.

Table 1: Average sulphur content using *n-* heptane solvent and microwave irradiation

Sample Identity	Sulphur Content (wt%)	Percentage Reduction (%)
Untreated Crude Oil	1.494	-
Oxidized Crude Oil	1.117	25.23
EOSC- 1:1 (Heptane)	0.273	81.73
EOSC- 1:2 (Heptane)	0.217	85.47
EOSC- 1:3 (Heptane)	0.318	78.79
Mw (300W-10mins)	0.692	53.68
Mw (300W-15mins)	0.322	78.45

Table 2 presents the average sulphur compounds in the untreated, oxidized and sulphur extracted crude oil with methanol solvent. On comparison, about 24.03, 31.12 and 30.72% sulphur was removed by using 1:1, 1:2 and 1:3 ratios; whereas microwave treated at 300W for 10 and 15minutes gave 53.68 and 78.45% sulphur reduction. These prove that methanol extraction is less effective in sulphur removal than *n-* heptane as well as being microwave irradiated.

Table 2: Average sulphur content using methanol solvent and microwave irradiation

Sample Identity	Sulphur Content (wt%)	Percentage Reduction (%)
Untreated Crude Oil	1.494	-
Oxidized Crude Oil	1.117	25.23
EOSC- 1:1 (Methanol)	1.135	24.03
EOSC- 1:2 (Methanol)	1.029	31.12
EOSC- 1:3 (Methanol)	1.035	30.72
Mw (300W-10mins)	0.692	53.68
Mw (300W-15mins)	0.322	78.45

Figure 1 presents a spectrum of untreated oil. A very broad IR absorption at 3445cm^{-1} centered between 3328-3448cm^{-1} was not used; hence it corresponds to –OH stretching vibration which may be attributable to intermolecular hydrogen bonding from water molecules, H$_2$O in the crude oil. Most of all spectra are similar but band 3445cm^{-1} those not appeared in spectrum of oxidized and irradiated crude oils.

Antonialli *et al.* [21] observed that aliphatic saturated C-H stretching bands occur within 2850 - 3000cm^{-1} range with intensities proportional to the number of C–H bonds. However, a strong infrared bands at 2924 and 2860cm^{-1} are responsible for methylene C-H and methyl –CH$_3$ asymmetric stretching vibration from saturated aliphatic compounds. Peak located at 1715cm^{-1} in the spectrum of untreated crude oil is carbonyl group, C=O due to connection to an aromatic ring [22]. IR absorption bands at 1457 and 1603cm^{-1} are typical of methyl, C–CH$_3$ asymmetric bend stretching from aromatic compound and carbon -carbon, C–C stretching showing skeletal vibrations on the ring.

John [22] observed in a practical approach that molecules containing nitro groups,-NO$_2$ exhibits vibration at 1260-1390cm^{-1}. It worthy to note that peak at 1375cm^{-1} and bands at 1302cm^{-1} are due to aliphatic nitro compounds, -NO$_2$ or nitrate ion. Absorption at 811 and 732cm^{-1} are due to 1,4 disubstitution (*para*) on aromatic determined from C–H vibrations and methylene, (CH$_2$)n rocking from aliphatic saturated compound. These observation, indicate that sample crude oil composed largely paraffinic and aromatic.

It is worthwhile to mention that sulphur and its compound structural information and identification is our priority in this study. Weak absorptions at 1032cm^{-1} is assigned to anhydride responsible for overlapping vibration of C–O–C stretching of ether and S=O of sulphoxide. The weak band at 617cm^{-1} in the untreated crude spectrum is assigned to disulphide, S–S stretching vibration due to catenation (unique chemical characteristic) where the formation of S–S bonds in extended C–C chain is common. Indeed, Further study by Al-Zahrani [18] revealed that oxidation process using oxidants, H$_2$O$_2$ convert sulphur compounds and there methyl and high alkyl derivative into sulphone which are removed by solvent extraction as sulphoxide. The peak located at 1017cm^{-1} is due to anhydride (ethers), C-O-C. However, focusing on sulphur compound identified, the strength of the absorption (or transmittance) is proportional to the concentration. These bands indicate the presence of sulphur and its group in the crude oil.

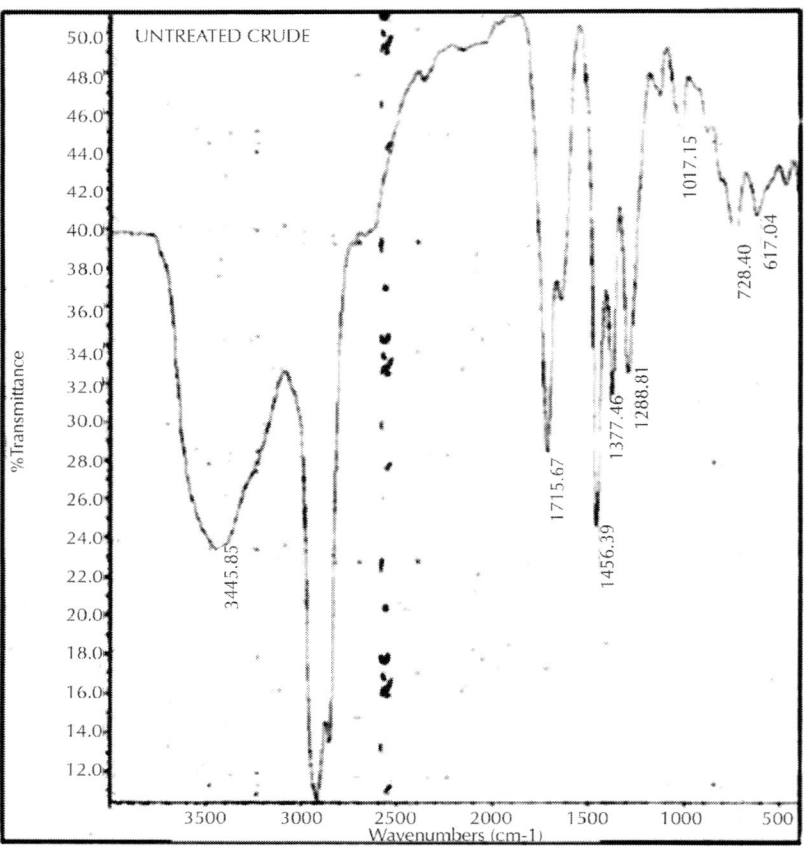

Figure 1: FT-IR spectra of untreated heavy crude oil.

Figure 2 presents the spectra of oxidized crude oils using hydrogen perioxide, H_2O_2. A spectrum pattern of oxidized oil is similar to that observed in Figure 1. Absorption bands at 2924, 1457 and 1375cm^{-1} assigned to methylene, C-H, methyl, C-CH$_3$ bend and aliphatic nitro compounds, NO$_2$ vibrations appeared in the spectra of untreated and oxidized crude oil. Whereas, there are bands at 2860, 1603 and 733cm^{-1} which does not appeared in some spectrum. These peaks are due to methyl, CH_3, carbon- carbon, C-C stretching exhibiting skeletal vibration on the ring. The weak band located at 617cm⁻¹ assigned to disulphide, S–S stretching vibration does not appear completely in treated samples spectra. This indicates effectiveness of oxidation process with low sulphoxide formation based on transmittance, 69.74%T at 1029cm⁻¹ assigned to sulphoxide.

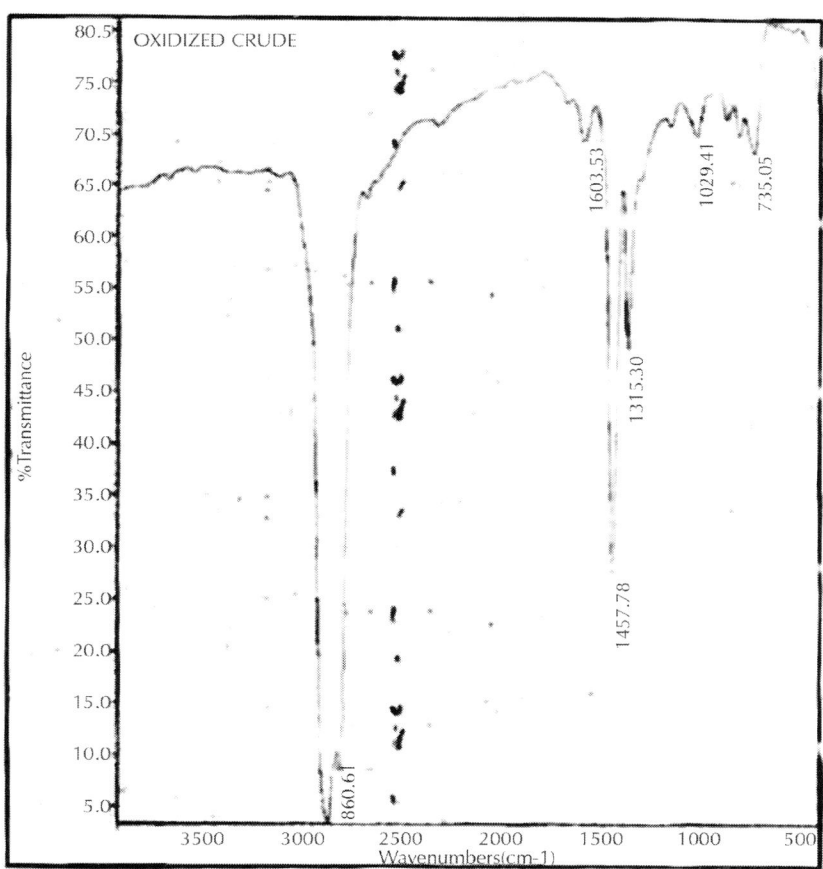

Figure 2: FT-IR spectrum of oxidized heavy crude oil

Figure 3(a) and (b) depict the spectra of irradiated at 300W for 10 and 15minutes. It is obvious the spectra are similar and showed the same pattern. Absorption bands at 2924, 2860 and 2359cm⁻¹ were assigned to methylene, C-H, methyl, C-CH$_3$ bend and carbon dioxide; CO$_2$ compounds vibrations by [22]. Peaks located at 2860 and 2359cm^{-1} do not appeared in the spectra of untreated and irradiated crude for 10minutes. Whereas, bands at 1604, 1458 and 1375cm^{-1} could be assigned to carbon- carbon, C-C stretching exhibiting skeletal vibration on the ring, methyl, C-CH$_3$, and nitro, NO$_2$ compounds in the asphaltenes fraction of the oil.

In view of the weak- to- moderate bands associated with sulphur bonds as seen at absorption band located 1033cm^{-1} do not appeared in the untreated crude oil and irradiated crude for 15minutes spectra but it is obvious in the oxidized and solvent treated crude by using 1:1, 1:2 and 1:3 ratio and irradiated crude for 10minutes spectra respectively. These are indicative of sulphide, S-S functional group oxidation in the presence of H$_2$O$_2$ to sulphoxide, S=O prior to solvent extraction. Frequency band at ~735cm^{-1} is due to methylene, (CH$_2$)n aliphatic long chain. In comparison, peaks located at 617cm^{-1} assigned to sulphides (S-S) seen not to appear in the solvent treated and microwave irradiated crude oil has it is been oxidized.

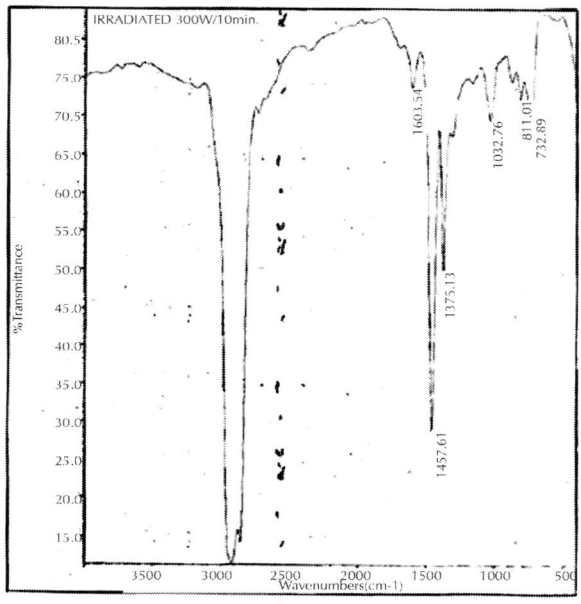

(a)

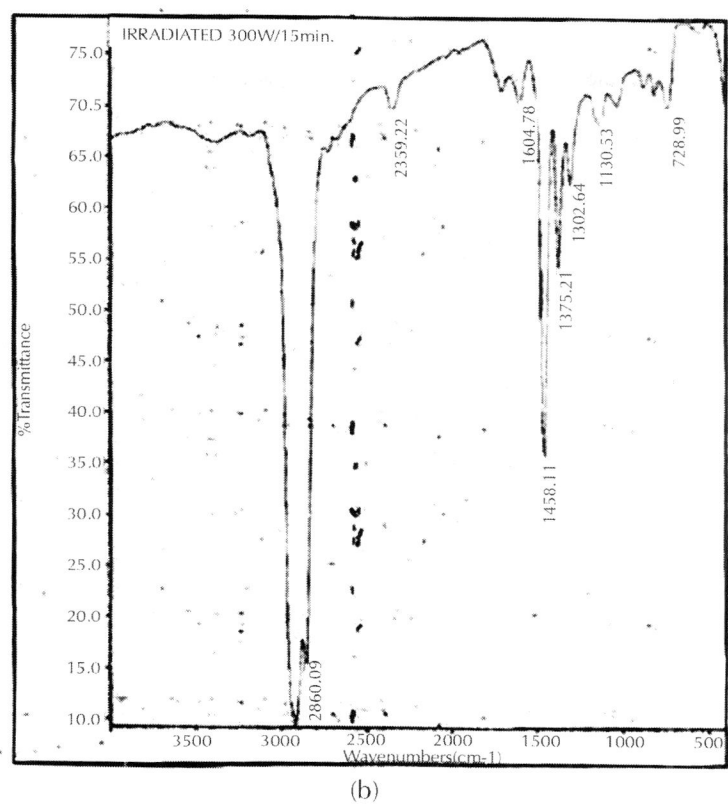

(b)

Figure 3: (a) FT-IR spectrum of irradiated heavy oil at 300W-10mins; (b) FT-IR spectrum of irradiated heavy oil at 300W-15mins

In Figure 4(a) the weak- to- moderate absorption band associated with sulphoxide bonds located at 1029 and 1033cm^{-1} does not appeared in the untreated crude oil spectrum but it is obvious in the oxidized and n- hexane solvent treated crude by using 1:1, 1:2 and 1:3 ratio. These are indicative of sulphide, S-S functional group oxidation in the presence of H_2O_2 to sulphoxide, S=O and anhydrate, C-O-C had occurred prior to solvent extraction. Interestingly, peaks located at 617cm^{-1} seen not to appear in the solvent treated crude oil due to its oxidation. Another important observation, the intensity of peak at 1029cm^{-1} assigned to sulphoxide using n- heptane extraction by using 1:1, 1:2 and 1:3 ratio are 83.52, 67.11 and 73.71%T respectively. It is obvious that extraction using 1:1 and 1:3 ratios gave high percentage sulphur reduction seen as sulphoxide compared to sulpur concentration in the oxidized crude oil. These results show that the solvent n- heptane have a high capacity to reduce the sulpur- containing compounds in the crude oil to a lesser content. It is quite clear that FT-IR analytical tool cannot determine the total sulphur content in weight percent of the oil rather their percentage concentration.

Figure 4(b) similarly presents spectra of desulphurized crude oils using methanol solvent extraction by 1:1, 1:2 and 1:3 ratio. Absorption bands located at 2924, 1603, 1456, 1375, 1033 and 735cm⁻¹ due to methylene, C-H, carbon- carbon, C-C, backbone skeletal stretch, methyl, $C-CH_3$, nitro, NO_2, anhydride, C-O-C and sulphoxide, S=O and methylene, $(CH_2)n$ aliphatic long chain were observed. In contrast, bands at 1304 and 1131cm⁻¹ due to aliphatic nitro compound, NO_2 in the aspaltene farction of the oil and secondary amine, C-N stretching.

However, comparing the spectra outputs in Figure 4(a) and (b) with untreated crude spectrum (Figure 1), it is observed that n- heptane and methanol solvents reduced the total sulphur respectively. Going by the intensity of absorption, methanol solvent on the average using extraction by 1:1 and 1:3 ratios proves less sulphur content reduction than using n- heptane. It is worthwhile clear that n- heptane solvent and microwave irradiation are suitable for the extraction of sulphur in the crude oil than methanol solvents.

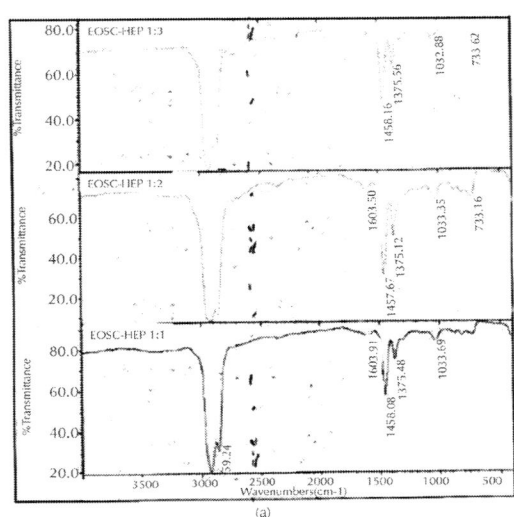

(a)

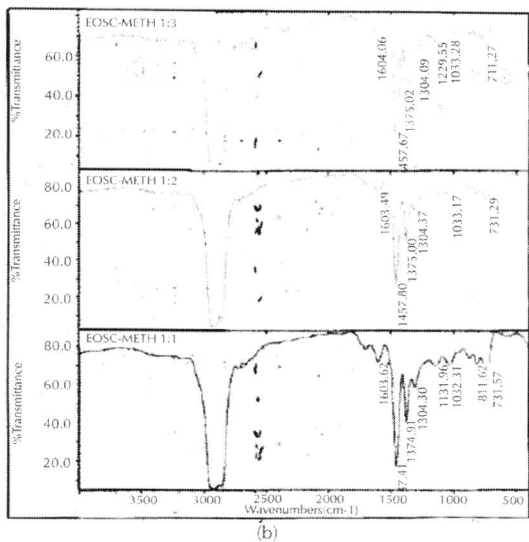

(b)

Figure 4: (a) FT-IR spectra of desulphurized crude oil with *n*- heptane at different ratio; (b) FT-IR spectra of desulphurized crude oil with methanol at different ratio

CONCLUSIONS

Sulphur reduction has been studied using microwave irradiation at 300W for 10 and 15minutes, *n*heptane and methanol solvents extraction by 1:1, 1:2 and 1:3 ratios respectively. The sulphoxide is the oxidized form of sulphur compound found in the crude oil as sulphides. Sulphur removal with *n*- heptane solvent by 1:1 and 1:2 are 81.73 and 85.47%; whereas when irradiated at for 10 and 15minutes, 53.68 and 78.45% sulphur reduction was achieved respectively. Comparing extraction using methanol by different ratios gave less sulphur reduction (24.03-31.12)% range to what was achieved under microwave treatments. Due to fragmentation of high molecular weight hydrocarbons caused by microwave irradiation in addition to air oxidation results to formation of alkyl radicals and sulphoxide from sulphides. In view of microwave results, 300W for 15minutes gave a remarkable sulphur reduction; this indicates microwave irradiation power rate is another factor that influence sulphur removal. The prevailing sulphur found in the crude going by FT-IR results is sulphides which oxidized to sulphoxide or sulpones. It is clear that *n*- heptane is more efficient than microwave radiation for sulphur removal but economically microwave treatment will be encouraged. However, on economic consideration, hence demands for *n*- heptane solvent is increasing as it also required, microwave irradiation can serve as alternative substitute in the extraction of sulphur from petroleum.

ACKNOWLEDGMENTS

The author appreciates the effort of Mall. Abdul-hameed Muhammad of Nuhu Bamali Polytechnic Zaria, Nigeria and also would like to thanks Laboratory assistance Mr. Sidi Ali of Chemical Engineering Department, FUT, Minna for his assistance during the studies.

REFERENCES

1.   Hanni V. D., Mustafa H.D. Innovative refining technology-Crude oil quality improvement (COQI). Real Innovators Group, Chemical Engineering Division, Santaeruz E, Mumbai, India, 2004, pp: 330- 335.

2.   Abdullahi D. M., Mohammed I. A., Ajinomoh C. S. Effect of crude oil preheating with microwave irradiation on distillates yield and propreties. M.Sc Thesis, Ahmadu Bello University, Zaria, Nigeria, 2010.

3.   Erikh V.N., Rasina M. G., Rudin M. G. The Chemistry and technology of oil and gas. Moscow, Mir publishers, 1984.

4.   Adebayo E. A. Introduction to petroleum processing. 1st edition, Clemol Publishers, Kaduna, 1999.

5.   OSHA Technical Manual. NW Washington, DC: 20210. Section IV: chapter 2. Available at: http://www.osha.gov/index.html, accessed in March 13, 2008.

6.   Nicholas E. Leadbeater, Rashid M. Khan. Microwave- promoted desulphurization of heavy and sulphur- containing crude oil. Energy & Fuels. American Chemical Society, xxx (xx): XXXX, AD, 2008.

7. Phuntdhawong W., Buddhasukh D. J., Pyre S., Rujiwatra A. Pakawatchai C. Microwave- assisted facile synthesis and crystal structure of cis- 9, 10, 11, 15- tetra- hydro- 9, 10 [3, 4]- furanoanthraceene- 12, 14- diona. Synth. Com. 2006, 39, 881.

8. Zhang M., Tang J., Mujumder A. S., Wang S. Trends in microwave- related drying of food and vegetables. Food Sci. Tech. 2006, 17, 524- 534.

9. Jackson M. Sulfur content of crude oils. Manual of sulphur content. Washington, U.S Department, 2002.

10. Ganjidoust H., Naghizadeh G. Interaction's effect of organic materials and aggregation on extraction efficiency of TPHS from petroleum contaminated soils with microwave assisted extraction. Iran J. Environ and Health Science Engineering. 2005, 2(4), 213- 220.

11. Camel V. Microwave- assisted solvent extraction of environmental samples. Trends in analytical chemistry. 2000, 19(4), 229 - 47.

12. Metaxas A. C., Meredith R. J. Industrial microwave heating. IEE. 2005, Reprinted 1988, 1993 and 1995.

13. Petr K., Hajek M., Cirkva V. The electrode discharge lamp: A prospective tool for photochemistry part 3. The microwave photochemical reactor. J. Photochemistry and Photobiology A: Chemistry 140. Elsevier, 2001, 185- 189.

14. Letellier M., Budzinski H., Charrier L., Capes S., Dorth A. M. Optimization by factorial design of focused microwave assisted extraction of polycyclic aromatic hydrocarbons from marine sediment. J. Analytical Chemistry. 1999, 364, 228- 37.

15. Ruren X., Wenqin P., Jihang Y., Qisheng H., Jiesheng C. Chemistry of zeolites and related porous materials- synthesis and structure. John Wiley & Sons, Pet Ltd, Asian, 2007.

16. Zuloaga O., Etxebarria N., Fernandez L. A., Madariaga J. M. Optimization and comparison of microwave assisted extraction and soxhlet extraction for determination of polychlorinated biphenyls in soil samples using an experimental design approach. Talanta, 1999, 50, 345- 57.

17. Brian K. W. Fourier transform infrared spectroscopy. Available at: http://fourier_transform_infrared_spectroscopy, accessed on August 19, 2009.

18. Al- zahrani I. M. Model system desulfurization by using liquid-liquid extraction and adsorption techniques. M.Sc Thesis, King Fahd University of Petroleum & Minerals, Dhahran, Saudi Arabia, 2009.

19. Sudipa M. K., Oliver C. M., Corie Y. R., Courtney P. Sulphur characterization in asphaltene, resin, and oil fractions of two crude oils. Rose- Hulman Institute of Technology, Schlumberger Doll Research, Albert Einsten University, New York, 1996, 763-767.

20. Cemends N. N. Chemical kinetics and reactions. M. Russian Academic Series, 1984, 1(2), 218.

21. Antonialli Junior W. F., Súarez Y.R., Izida T., Andrade L.H.C., Lima S.M. Intra- and interspecific variation of cuticular hydrocarbon composition in two ectatomma species (Hymenoptera: Formicidae) based on fourier transform infrared photoacoustic spectroscopy. J. Genetic Molecular Research, 2008, 7(2), 559-566.

22. John C. Interpretation of infrared spectra: A practical approach in encyclopedia of analytical chemistry. R. A. Meyers (Ed.), John wiley & Sons, Newtown, USA, 2000, 10815- 10837.

Asphaltene Flocculation Inhibition with Ultrasonic Wave Radiation: A Detailed Experimental Study of the Governing Mechanisms

Iman Najafi[1]; Mahmood Amani[1,*]

[1]Texas A&M University at Qatar

ABSTRACT

The concept of ultrasonic wave assisted asphaltene flocculation/ deposition inhibition was earlier introduced by Najafi et al., (2011). Current study is based on a series of experimental analyses, rheological changes, flocculation behavior, and total asphaltene content of two types of crude oils. The role of ultrasonic wave radiation on these parameters are investigated in order to elaborate the changes in the kinetics of asphaltene flocculation which leads to inhibition of asphaltene flocs formation. Based on the results obtained from the experiments, one can conclude that ultrasonic wave radiation can be most effective if the wave is radiated up to an optimum time. Around this time the oil has its local minimum value of kinematic viscosity, the least value of asphaltene content, and least potential for formation of macro-structured flocsdue to reduction of aromatic to saturate ratio.

More detailed studiesrevealed that during radiation time two main mechanisms are active in asphaltenic crude oil: asphaltene particles disintegration and formation of asphaltene particles. Based on asphaltene content analysis, the optimum radiation time can be defined as the time at which the two mechanisms have an equal rate. The optimum radiation times are observed to be in the same time range in different tests. According to the results obtained, ultrasonic wave technology can be a potential method of flocculation inhibition and can have extensive industrial application.

Asphaltenes are heavy complex molecules which are soluble in aromatics and non-soluble in paraffins. They are the heaviest and the most polar fraction ofcrude oil. In petroleum chemistry it is predominantly accepted that asphaltenes are suspended in micelle form in petroleum and are stabilized by adsorbed resins in solutions (Gollapudi, 1994). The ratios of resin to asphaltene and aromatic to saturate are the key parameters that control the stability of asphaltene micelles in crude oil. When these ratios decrease, asphaltene micelles flocculate and form larger aggregates (Diallo et al., 2000). Injection of solvents into petroleum reservoirs in tertiary recovery methods can destabilize the micelle by stripping the resins from around the asphaltene molecules and would cause asphaltene depositions which will lead to drastic operation and economic complications in production, transport, and refining of crude oil.

In an earlier report (Najafi et al., 2011), the potential benefits of ultrasonic wave technology as a novel method of asphaltene flocculation/deposition inhibition was presented. This study is devoted to providing more information about the governing mechanisms in this method based on quantitative analyses.

In recent years, acousticwave technology's application to removing asphaltene deposits from near wellbore regions, cracking asphaltene molecules, etc. has been studied by many researchers. Champion et al. (2004) generated high power sound with a high voltage electrical discharge to investigate the applicability of acoustic waves for wellbore cleaning. They concluded that high power sound wave technology could be an effective method of removing wellbore plugging materials such as asphaltene. Gunel and Islam (2000) compared electromagnetic and ultrasonicwave's role on crude oil properties alteration. Their experiments showed that in the case of asphaltenic crude oil, ultrasonic waves can change the rheological properties of oil samples but these alterations are not long-lasting. The alterations made by electromagnetic waves are reported to last longer.

Dunn and Yen (2004) focused on the influence of ultrasonic waves on conversion of asphaltene molecules and concluded that sonication would lead to both dehydrogenation and cracking in bitumen. In 2009,Sawarkar et al. reported that a reduction in asphaltene content was observed due to conversion of refinery residues to lighter hydrocarbons in the boiling range of gasoil fractions. The reaction time in their experiments varied in a range of 15 to 120 minutes.

Shedid and Attallah (2004) focused on the influence of ultrasonic waves on rheological behavior of UAE crude oils in different solvent concentrations. Temperature and solvent concentration's role were investigated in series of experiments. Microscopic studies and differential thermal analyses were carried out to analyze the experimental results. Based on their work, ultrasonic wave radiation decreased the size of asphaltene flocs. This will reduce / prevent precipitation at 10 minutes of radiation or more. In their study, asphaltene content of oil samples were 1.76 weights percent.It seems that using crude oil with more asphaltene contents would lead to more apparent results. Based on a series of crude oil rheological properties and asphaltene flocculation confocal microscopy analysis, Najafi et al. (2011) reported the existence of an optimum radiation time at which asphaltenic crude oils reach the minimum kinematic viscosity. Experiments on asphaltene flocculation process in toluene-n¬-pentane mixtures showed that wave radiation can change both flocculation rate and flocs size distribution. Accordingly, they proposed the idea of asphaltene flocculation inhibition due to wave radiation.

THE PROCESS OF FLOCCULATION INHIBITION

The present investigation is a continuous effort to provide more information about the process of flocculation inhibition. Confocal microscopy and rheological analyses were performed on different crude oils to prove the repeatability of the observed phenomena. Asphaltene content analysis was done based on IP143 procedure, which will lead to more viable conclusions in this study.

If it is proved that ultrasonic waves can reduce the rate of flocculation and the tendency of asphaltene particles to floc, many of the production obstacles will be removed. The main focus in the next parts of this paper will be on the role of ultrasonic waves on flocculation of asphaltene particles due to reduction of aromatic to saturate ratios in solvent injection operations.

When a wave is radiated to a liquid it will cause cavitation and consequently some micro-streams form in the liquid environment. These two mechanisms can break down and resolve the asphaltene molecular structures in crude oil. This will lead to a change in crude oil composition which in turn results in changes in the rheological behavior of crude oil. To quantify the changes a series of viscometery tests are performed.

A Cannon-Fenske Routine Viscometer-100 was used to measure the viscosity variation of crude oil samples at different time intervals of ultrasonic radiation. As the asphaltene particles are colloidally suspended into the crude oil and there are no two separate solid and liquid phases, this type of viscometers seem to report accurate values. To be more confident about the results, the measurements were performed more than one time. The ultrasonic radiation was applied using a URG500 Wave Generator with 45-kHz frequency and 75-Watts output power. The value of output power is not the nominal output of the system and was calculated based on calorimeter experiments. To know more about the calorimetery method used, refer to Kikuchi and Uchida (2011).The proper power rating was achieved to mix the crude oil well. Two different crude oils were used in these experiments. The initial asphaltene content of the crude oils L and H was measured to be 10.2wt% and 12.3 wt% using IP-143 method. Toleune and n-heptane were used in this method. The API gravity of each of the crudes was reported to be 20 and 12 degrees, respectively. After the ultrasonic radiation to 200 ml of each crude oil for the specified time intervals, the oil samples were cooled to ambient temperature and the measurement was taken.

Based on previous studies, it can be inferred that ultrasonic wave can play three main roles in crude oil, (Suslick, 1989)

1. Dissolution of suspended soluble particles in crude oil, and
2. Increasing the temperature of crude oil,
3. Disintegration/formation of long chain molecules and asphaltene flocs.

The resultant of these three roles in this study leads to a total increase in crude oil viscosity in both samples. As is shown in Figures1 and 2 there is an optimum radiation time at which the value of viscosity has a local minimum. This time was recorded to be between 10 to 25 minutes for sample L and between 30 to 50 minutes for sample H.

In general three ranges could be determined by these figures, as is shown in Figures 1 and 2, at each range one or two of the mechanisms are dominant. Previously Argillier et al. (2002) mentioned the concept of critical concentration of asphaltenes, above which the overlap made between asphaltene particles would lead to an increase in viscosity. The rheological behaviors at each of these ranges can be justified based on the influence of ultrasonic waves on crude oil, as follows.

Range 1: In this range the main role is played by dissolution of unsolved suspended particles in crude oil which leads to an increase in viscosity.

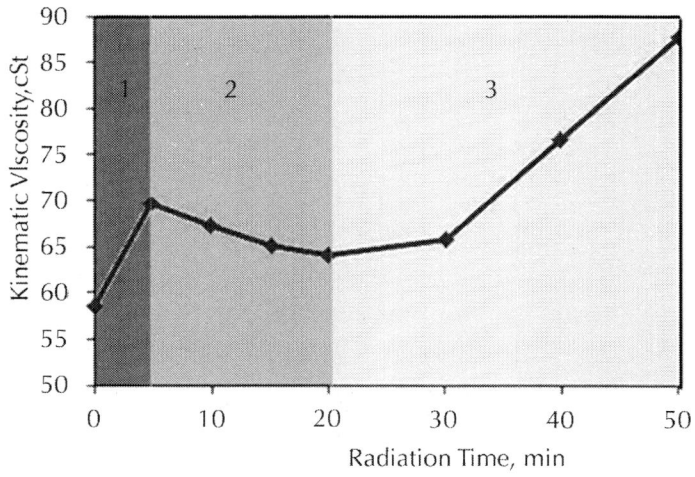

Figure 1: Viscosity Changes versus Radiation Time for Crude Sample L

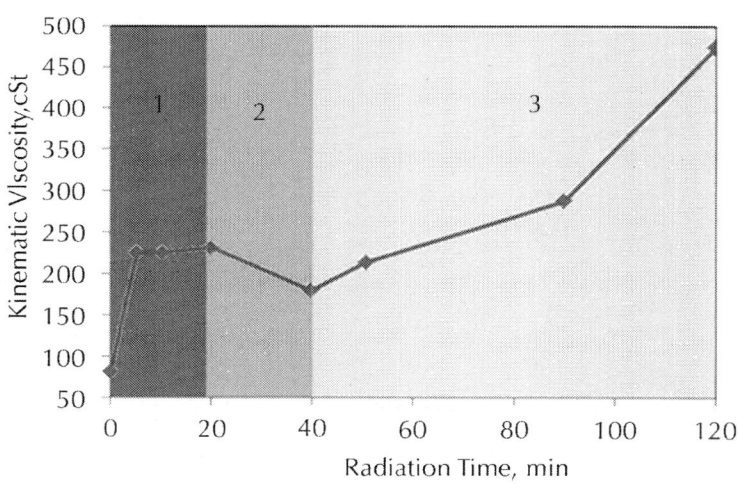

Figure 2: Viscosity Changes versus Radiation Time for Crude Sample H

Range 2: The main reasons for viscosity reduction in this range are temperature increase and asphaltene molecules disintegration. Ultrasonic wave cracks the long chain and heavy molecules and forms free radicals which leads to a reduction in viscosity.

Range 3: In all periods of sonication two mechanisms are active in crude oil system. These mechanisms are molecules disintegration into free radicals and hydrocarbons with shorter chain length, and integration of free radicals. Respectively these mechanisms cause formation of free radicals and formation of heavy molecules. This means that sonication can change the composition of crude oil. Based on the viscosity curves on Figures1 and 2,it is deduced that after the local minimum point, namedthe optimum radiation time, the dominant factor is reintegration of free radicals to each other. The next part of this paper is devoted to verify these justifications.

Comparing Figures 1 and 2 shows that an increase in API will result in an increase in optimum radiation time.

For Sample H the optimum time is at about 40 minutes while it appears to be in the range of 10 to 20 minutes for Sample L.

If sonication can change the composition of a crude oil, kinetics of asphaltenic particles flocculation for crude oil samples radiated at different periods of time are anticipated to be diverse. To prove this, confocal microscopy experiment were performed. The procedure was the same as that in Najafi et al. (2011).

To study flocculation kinetics of asphaltene particles, 15 ml of n-pentane was added to 10 ml of 5% crude oil in toluene mixture; this ratio of crude oil/n-pentane ensures the flock formation. At selected times of flocculation 3 drops of the sample were taken to be observed in the confocal microscope. The light source consisted of a Tungsten lamp. The samples were observed with a magnification of 500. The microscope was connected to a PC and images of the flocculated asphaltenes were stored in raw format of 640×480 pixels. To obtain a statistically reliable particles size distribution curve, more than 10 images were taken per sample. More than 10000 images were stored to analyze the size as a function of time. Each value representing the size of flocks is an average of more than 150 particles in each sample. Photoshop software was used to determine the size of asphaltene flocks in different samples.

As it is shown in Figure 3the radiation of ultrasonic waves changed the irreversible mechanism of flocculation to a reversible phenomenon in both samples. This means that the asphaltene particles in the sonicated crude oil and n-pentane mixture have a greater tendency to separate and disintegrate flocs, rather than forming new flocs. The phenomenon was not observed in the same crude oil which is not sonicated. This shows that although irreversible models of flocculation like DLVO models can predict the flocculation behavior of asphaltene particles in non-sonicated crude oils, these models cannot be applied to the sonicated samples of the same crude. Figure 4 shows that for crude Samples L and H the least value of average flocsradius and rate of flocs growth are for Sample L3 and Sample H3.These samples were radiated for 10 minutes and 40 minutes respectively. These figures are showing the period of flocculation time at which the flocs formation rate is more than the disintegration rate.

Samples L1 to L4 were radiated for 0, 5, 10, and 20 minutes, respectively. The histograms of asphaltene flocs size at different flocculation times are depicted in Figure 5. An observed shift of the diagram to the left hand side could be due to radiation of ultrasonic wave. This means that at a specified time of flocculation the crude oil which was treated with ultrasonic wave has less potential for generating macrostructure flocs.

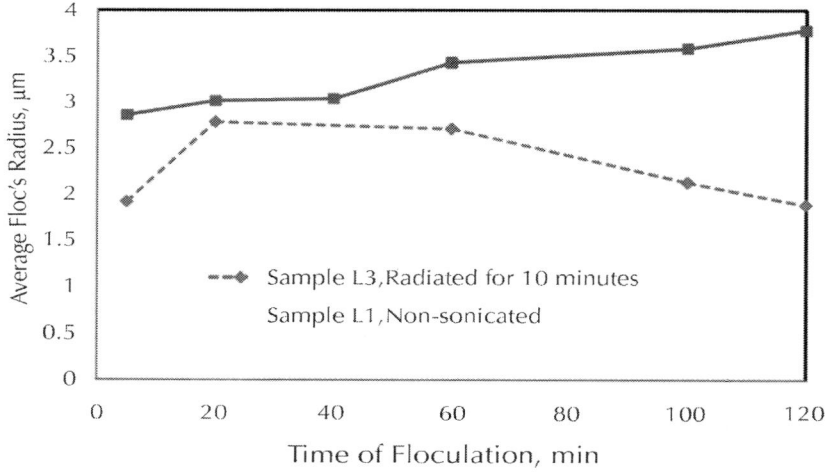

Figure 3: A Sphaltene Flocs Average Radius in Crude Sample L in Different Flocculation Times

It is known that flocs with lower sizes have a lower tendency to precipitate. As is shown in Figure 5, the trend of shifting to the left hand side, i.e. smaller flocs, is observed for the first three samples of crude oils. The third samples are the ones which are radiated for the optimum time. It was shown that the fourth sample had more tendencies for formation of bigger flocs than third sample. Based on these histograms, it can be concluded that before optimum time, radiation decreases the potential of asphaltenes for being precipitated while after this time the trend changes. It seems that radiation can make the asphaltene molecules smaller, by breaking down the chemical bonds until optimum radiation time. The result of this would be formation of free radicals which can rejoin by collision together. As the radiation time increases, these radicals' formation rate will decrease and reintegration rate will increase. After optimum time, it seems that reintegration of free radicals led to formation of heavier components and the asphaltene molecules with more branches.

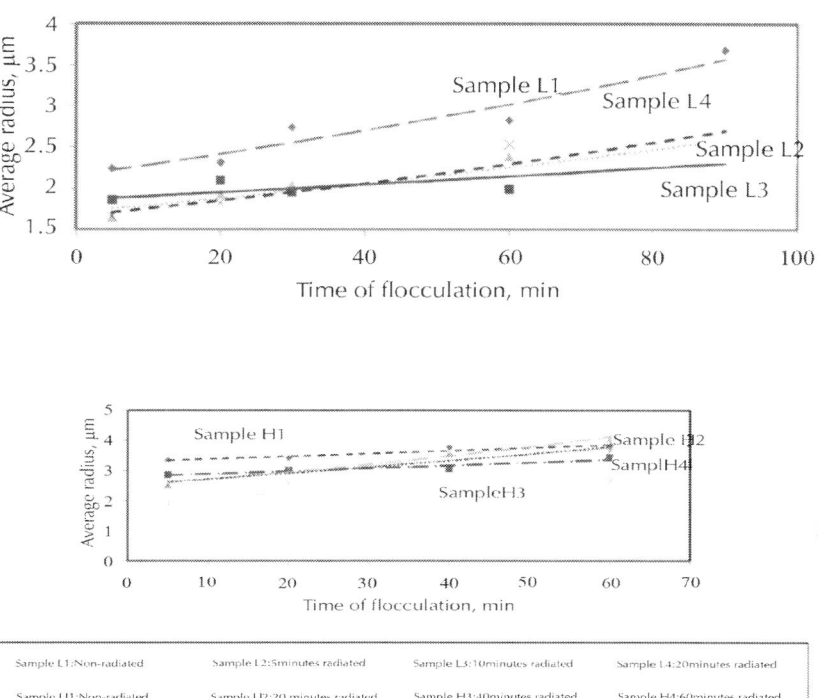

Figure 4: Trend of Asphaltene Particles Radius Increase versus Time of Flocculation in Presence of Alkane for L and H Crude Oils

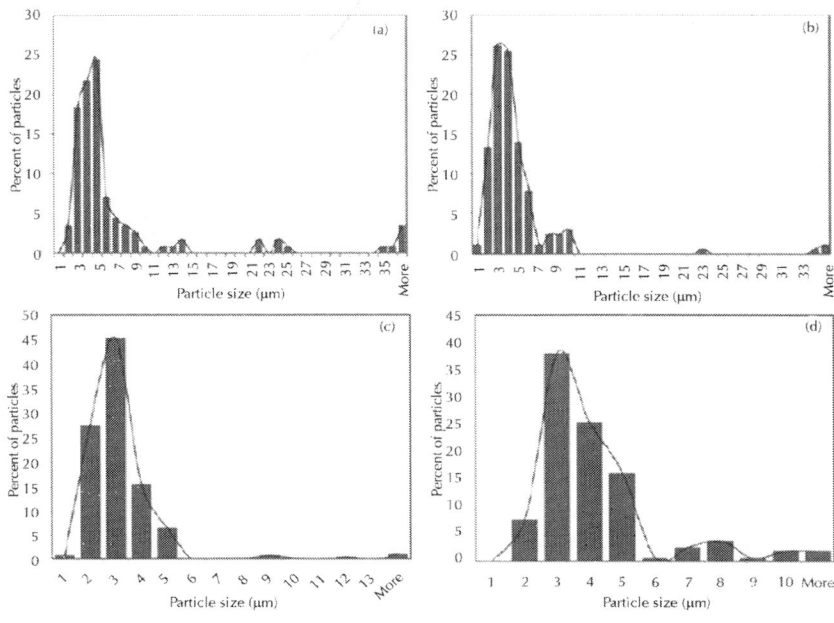

Figure 5: Particles Size Distribution Curve for Four Samples L Crude Oil Flocculated with 60% n-Pentane after 120 Minutes of Flocculation. (a) Sample L1, (b) Sample L2, (c) Sample L3, (d) Sample L4

In order to quantify the changes in asphaltene content of crude oil in a direct way a series of IP143 tests were done. In these tests the total asphaltene content of crude oils were measured. The radiation time to each sample was selected based on the results of previous experiments. Crude Sample Lwas radiated for 5, 10, 15, 20 and 25 minutes and Sample H was radiated for 10, 20 , 40, 50 and 60 minutes. For measuring asphaltene content of crude oil IP143 procedure with n-heptane and toluene as solvents, was applied.

Based on our previous experiment it was shown that ultrasonic wave radiation for a specified time can reduce asphaltene content of crude oil. As is shown in Figures 6 and 7, this reduction is measured to be about 11% for sample L and 27% for sample H. It proves that before optimum radiation time ultrasonic could break down the asphaltene particles. Focusing on Figure 7 one can see that the slope of the curve reduces and tends to zero. At the optimum point, the slope is zero and it means that the rate of asphaltene particles disintegration is equal to the rate of asphaltene formation. After this point, asphaltene formation process is dominant.

The results obtained by IP143 tests are in accordance with the results of the confocal microscopy and viscosity tests.

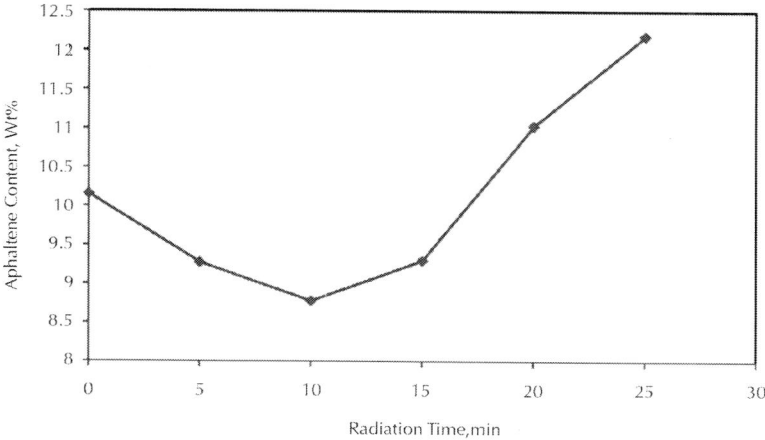

Figure 6: Asphaltene Content of Sample L Crude Oil Radiated for Different Times

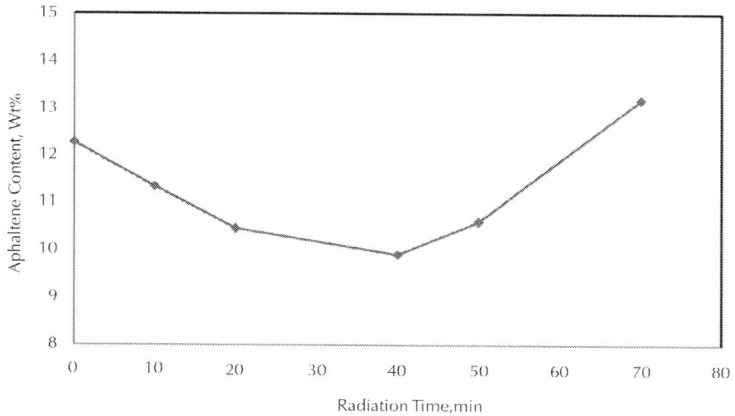

Figure 7: Asphaltene Content of Sample H Crude Oil Radiated for Different Times

CONCLUSIONS

In this study, viscometery tests, confocal microscopy and IP143 asphaltene content analysis are performed to justify the role of sonication on asphaltene flocculation in two samples of asphaltenic crude oils and the following conclusions are made:

1. Two mechanisms are always active while ultrasonic radiation, which are asphaltene particles disintegration and rejoining of free radical to form new asphaltene particles. Optimum radiation time is the time at which the rate of formation is equal to the rate of disintegration. Before optimum radiation time disintegration and after that formation is dominant.

2. It seems that heavier crude oils will have greater optimum radiation time. This time was measured to be 10 minutes for sample L and 40 minutes for sample H. Optimum radiation time measured was in accordance together in different tests.

3. It seems that performing a successful ultrasonic radiation operation requires a good knowledge of the fluid characteristics of the well and the optimum condition varies from well to well.

ACKNOWLEDGMENT

The authors would like to express their appreciation for **Times** the contributions and guidance provided by Dr. Cirus Ghotbi, Dr. Mohammad Hosein Ghazanfari and Mr. Mohammad Reza Mousavi.

REFERENCES

1. Argillier, J. F., Coustet, C., & Hénaut, I., (2002). *Heavy Oil Rheology as a Function of Asphaltene and Resin Content and Temperature.* SPE/Petroleum Society of CIM/CHOA 79496, International Thermal Operations and Heavy oil Symposuim and International Horizontal Well Technology Conference.

2. Sawarkar A. N., Aniruddha B. P., Shriniwas D. S. & Jyeshtharaj B. J. (2009). Use of Ultrasound in Petroleum Residue Upgradation. *The Canadian Journal of Chemical Engineering, 87,* 329-342.

3. Champion, B., Van der Bas, F., & Nitters, G. (2004). *The Application of High-Power Sound Waves for Wellbore Cleaning.* The Hague, Netherlands: SPE 82197, SPE European Formation Damage Conference.

4. Diallo, M. S., Cagin, T., Faulon, J. L., & Goddard W.A. (2000). Thermodynamic Properties of Asphaltenes: A Predictive Approach Based on Computer Assisted Structure Elucidation and Atomistic Simulations, Asphaltenes and Asphalts, 2. *Developments in Petroleum Science, 40*(B), 103-127.

5. Dunn, K., & Yen, T. Y. (2001). A Plausible Reaction Pathway of Asphaltene under Ultrasound. *Fuel Processing Technology, 73,* 59–71.

6. Gunel, G. O., & Islam, M. R. (2000). Alteration of Asphaltic Crude Rheology with Electromagnetic and Ultrasonic Irradiation. *Journal of Petroleum Science and Engineering, 26,* 263–272.

7. Gollapudi, U. K., Bang, S. S., & Islam, M. R. (1994). *Ultrasonic Treatment for Removal of Asphaltene Deposits during Petroleum Production.* SPE 27377, SPE Intl. Symposium on Formation Damage Control, Lafayette, Louisiana, U.S.A.

8. Kikuchi, T., & Uchida, T. (2011). Calorimetric Method for Measuring High Ultrasonic Power Using Water as a Heating Material. *Journal of Physics: Conference Series, 279,* 1-5. doi:10.1088/1742-6596/279/1/012012.

9. Najafi, I., Mousavi, S. M. R., Ghazanfari M. H., Ramazani, A., Kharrat, R., Ghotbi, C., & Amani, M. (2011). *Quantifying the Role of Ultrasonic Wave Radiation on Kinetics of Asphaltene Aggregation in a Toluene-Pentane Mixture. Petroleum Science and Technology, 29*(9), 966-974.

10. Shedid, A. S., & Attallah, S. R. (2004). *Influences of Ultrasonic Radiation on Asphaltene Behavior With and Without Solvent Effects.* SPE 86473, SPE International Symposium and Exhibition on Formation Damage Control, Lafayette, Louisiana, U.S.A.

11. Suslick, K. S. (1989). The Chemical Effects of Ultrasound. *Scientific American, 260*(2), 80- 86.

Electromagnetic Heating of Heavy Oil and Bitumen: A Review of Experimental Studies and Field Applications

Albina Mukhametshina[1,2] and Elena Martynova[1,2]

[1]Harold Vance Department of Petroleum Engineering, Texas A&M University, 3116 TAMU-407 Richardson Building, College Station, TX, USA
[2]Gubkin Russian State University of Oil and Gas, 65 Leninsky Prospekt, Moscow, Russia

Received 11 December 2012; Revised 6 March 2013; Accepted 11 March 2013

ABSTRACT

Viscosity is a major obstacle in the recovery of low API gravity oil resources from heavy oil and bitumen reservoirs. While thermal recovery is usually considered the most effective method for lowering viscosity, for some reservoirs introducing heat with commonly implemented thermal methods is not recommended. For these types of reservoirs, electromagnetic heating is the recommended solution. Electromagnetic heating targets part of the reservoir instead of heating the bulk of the reservoir, which means that the targeted area can be heated up more effectively and with lower heat losses than with other thermal methods. Electromagnetic heating is still relatively new and is not widely used as an alternate or addition to traditional thermal recovery methods. However, studies are being conducted and new technologies proposed that could help increase its use. Therefore, the objective of this study is to investigate the recovery of heavy oil and bitumen reservoirs by electromagnetic heating through the review of existing laboratory studies and field trials.

INTRODUCTION

High-frequency electromagnetic radiation is a relatively new technique for use in enhanced oil recovery methods. It has been tested by theoretic, laboratories and field trial research in Russia [1–10], the United States [11–17], Canada [18–21], and other countries [22–34]. Traditional thermal recovery and well stimulation techniques using hot steam or fluid are not effective in some cases [7, 35] due to prohibitive heat losses from injection wells and reservoirs, low reservoir injectivity (especially for bitumen deposits), steam leakage, large overburden heat loss at thin pay zones, permafrost conditions, and so forth. Furthermore, commonly used thermal recovery methods are not considered environmentally friendly, damaging the hydrogeologic environment and contributing to the greenhouse effect.

The most important thing in electromagnetic heating is that the heat is developed within the material rather than being brought from outside, which means the material is heated more uniformly throughout the medium [27]. Therefore, instead of heating the bulk reservoir volume, part of the reservoir can be targeted and heated more effectively with lower heat loss than other thermal methods. Unlike traditional thermal recovery methods, microwave heating causes friction by vibration of molecules, which results in dielectric heating of the reservoir. Heat and mass transfer in different environments under microwave influence was studied by a number of scientists around the globe, but its application as an EOR method is not yet fully understood. Microwave heating is not used productively because of the lack of reliable information about the processes of heat and mass transfer in a multiphase system in porous media under the influence of electromagnetic radiation, which does not allow effective control. Therefore, current research studies use modeling to discover optimum design parameters for the use of microwave heating in field applications.

REVIEW OF EXPERIMENTAL ELEC-TROMAGNETIC HEAT STUDIES

The success of near wellbore heating with electromagnetic waves has been proven experimentally [1]. To represent reservoir rock, quartz sand with a 7.7 dielectric constant and a 0.083 tangent was used in the laboratory. A 20% initial water saturation and an 80% initial oil saturation with 16.61 cps ($16.61*10^{-3}$ Pa·s) gravity and 0.86 g/cm^3 (860 kg/m^3) density oil were maintained. The dielectric constant and the loss tangent of the oil sample were 2.23 and 0.019, respectively. In this study of electric and high-frequency electromagnetic heating of a reservoir model, the temperature of the medium was measured by thermometers located at different points of the experimental setup. In another case, a linear radiator with a length of 87 cm and a diameter of 19 mm was placed in the center of the setup. The linear radiator was connected via acoaxial cable to the generator supplying electromagnetic waves with a frequency of 13.56 MHz. In the experiment, it was discovered that when fluid temperature was exposed to a high-frequency electromagnetic field at the same distance from the radiation source it was greater than one initiated by electric heating. In this case, the thermal conductivity of the medium is affected only slightly. When heat is induced throughout the whole volume of the medium, the amount of heat introduced depends largely on the electrical properties of the medium. The study summarized the advantages of microwave heating over electrical heating as having deeper penetration, quicker heating, and lower heat losses.

Sayakhov [2] discussed the physical foundations of fluid filtration in high-frequency fields. Radial fluid filtration through porous media in a heterogeneous high-frequency electromagnetic field was studied experimentally. A specially designed core holder, filled with a representative porous medium, was located in a coaxial resonator. A high-frequency wave generator was used to generate mw energy, which was directed to the resonator through cables. Studied liquid came into the porous media through the hose first, then through the lumen of the inner conductor of the resonator dripped into a graduated cylinder. Experiments were started with a 200–300 watts power microwave oven at 2400 MHz frequency and 500 watts vibrational power. Temperatures were recorded throughout the experiments via thermocouples inserted in the core holder. Kerosene was used as a representative reservoir fluid. The measurements were performed with and without the influence of a high-frequency electromagnetic field. It was established that exposure to high-frequency electromagnetic fields leads to a sharp increase in flow rate per unit of time and fluid temperature at the outlet. In addition, the flow rate increases dramatically once exposed to the electromagnetic field, while the temperature increases after 10 seconds. After the field is discontinued, a sharp decrease in the flow rate is observed and a gradual cooling of the porous medium takes place.

Experimental studies [3] on the influence of electromagnetic fields with a frequency of $3*10^5$ Hz to $6*10^5$ Hz on thermal conductivity of dielectric liquids showed that the thermal conductivity of the liquids increased when exposed. The thermal conductivity increases as the magnitude of the dipole moment of the liquid used in the experiment parameters increases with the frequency and intensity of field.

Fatikhov [5] conducted experimental research on the flow of bitumen oil at different pressure gradients in a high-frequency electromagnetic field. The experiment focused on changes in the volumetric flow rate of filtered oil under different pressures at different temperatures. The initial pressure drop for bitumen from the Mordovo-Karmalskoye deposit in reservoir conditions was 0.003 MPa/m. During the experiment, the pressure drop decreased rapidly as the temperatures were increased, resulting in bituminous oil becoming a Newtonian liquid. Therefore, it was established that the application of electromagnetic heating improves fluid flow behavior and the non-Newtonian properties of bitumen decrease rapidly with increasing temperature.

In 1992, Kasevich et al. [17] studied electromagnetic heating of rock samples at 1 kW (frequency 50.55 MHz) and 200 W (144 MHz) and found that a particular type of rock could be heated to 423 K when exposed to an RF electromagnetic field. The rock, which had low thermal conductivity, heats poorly when hot steam is pumped into the reservoir. They conducted experiments at both naormal and formation pressures.

Ovalles et al. [30] used a microwave with 650 watts power to heat core samples saturated with oil of 25 API gravity and 7.7 API gravity (a sample from the Orinoco River Basin). Medium API oil temperatures were measured in 0.5, 1, and 1.5 minute intervals and heavy oil intervals were increased to 1, 5, and 10 minutes. The experimental results were used to test mathematical models and predict the production of three abstract oil reservoirs in Venezuela.

Chakma and Jha [23] conducted laboratory experiments using electromagnetic heating on a scaled thin heavy oil reservoir pay zone model. Gas injection with horizontal wells during electromagnetic heating was achieved. The aim was to decrease oil viscosity with electromagnetic heating and obtain a gas drive with the injected gas. Using nitrogen for the injected gas, they were able to prove that for thin pay zones heating of the wellbore vicinity is sufficient, by achieving oil recoveries as high as 45% of original oil in place compared to estimated primary recovery rates of less than 5%. Recovery achieved by use of the combined method was higher than that of nitrogen injection or electromagnetic heating alone. Chakma and Jha also discussed a number of parameters affecting the results of the combined method, including(i)gas injection pressure (when no gas was injected, oil was produced only due to gravity drainage and no significant convective transport occurred; therefore, gas injection provided an oil rate increase with the increasing injection pressure);(ii)temperature (the initial production rate was not significantly affected by temperature, but later there was an increase in the production rate, meaning that overall recovery increased with temperature);(iii)electromagnetic frequency (the higher the frequency, the greater the recovery);(iv)oil viscosity (as expected, a higher oil viscosity leads to a lower recovery for a given electromagnetic frequency, temperature, and gas injection pressure);(v)salinity (higher salinity provides higher recovery due to the higher conductivity of saline water compared with distilled water);(vi) electrode distance (recovery is similar, but closer electrode spacing provides faster production rates) [23].

Hascakir et al. [24] conducted a laboratory study of microwave-assisted gravity drainage on heavy oil samples from reservoirs in Turkey (Bati Raman, 9.5 API; Garzan, 12 API; and Camurlu, 18 API) using a specially designed novel graphite core holder packed with crushed limestone. Their study described the effects of operational parameters like heating time, waiting period and rock, and fluid properties on the effectiveness of microwave heating. Some of the conclusions made are confirming ones found in Chakma and Jha [23], like the positive effect of high water salinity and water saturation. Hascakir et al. [24] also concluded that water wet conditions are preferable for obtaining higher oil recoveries and that large porosity and permeability are also favorable. When microwave heating is applied to oil samples continuously, higher temperatures are reached, which allows better results to be achieved than with periodic heating when microwave heating is applied for a limited time in periodic intervals. Therefore, higher temperatures allow for better results in continuous heating.

Jha et al. [27] proposed using microwave-assisted gravity drainage (MWAGD) in the Mehsana oil field in India. They heated specially prepared samples with the required characteristics from that field in the laboratory using a microwave with variable power up to 1000 watts operating at 3 GHz frequency, which allowed them to obtain temperature and viscosity profiles of the gravity-drained oil. They described effects of initial oil and water saturations, wettability, porosity, and permeability similar to those found by Hascakir et al. [24] and Chakma and Jha [23].

Jha et al. [27] suggested using MWAGD commercially by drilling one horizontal well and multiple vertical ones with downhole microwave antennas. However, this might not allow deep enough heat penetration, so other options are also proposed such as a combination of two horizontal wells and installing antenna inside the horizontal production well. Vertical separation of the horizontal well pair is approximately 15 meters, which is far more efficient than SAGD in which the separation is around 5 meters. Because crude oil absorbs microwave heat weakly, Jha et al. also proposed increasing thermal conductivity by injecting powdered metallic oxides, chlorides, or activated carbon through a fracture operation. The working principle and description of laboratory applications of such additives to heavy oil can be found in various studies by Hascakir et al., Kershaw et al., and Odenbach [25, 28, 29].

Technical principles of the SAGD method assisted by electromagnetic heating (EM-SAGD process) were reported by Koolman et al. [26]. Inductive heating was initiated in the laboratory using an EM source with a working frequency of 142 kHz. The sample was heated for 10 minutes at a power of 7.2 kW, achieving a rise in the temperature of 7.5 K. Laboratory and field processes were modeled using a numerical simulator, combining electromagnetic and thermal modules. It was specially built and can be applied to field-scale simulations. According to simulation results, a 38% increase in bitumen production was predicted compared to conventional SAGD.

Kovaleva et al. investigated the effects of radio frequency electromagnetic (RF-EM) fields and electrical heating on the mass- and heat-transfer processes in a multicomponent hydrocarbon system flowing in porous media [4, 6]. Three different types of experiments were carried out: solvent (kerosene) flooding under the RF-EM field, solvent (kerosene) flooding under electrical heating, and cold solvent (kerosene) flooding. In all three experiments, the physical characteristics of the model and heating conditions (temperatures) were identically maintained. Two series of experiments on models with different granulometric composition of formation were also carried out.

Figure 1 shows the dependence of oil recovery on the volume of solvent injected. This figure demonstrates that the highest oil recovery was obtained by applying an RF-EM field. Kovaleva et al. [4, 6] concluded that following RF-EM influence on the oil-saturated samples the quantity of the received oil is more than the quantity received under electrical (and thermal) processing at identical temperatures of heating of the media. It confirms the additional "nonthermal" action of the electromagnetic field.

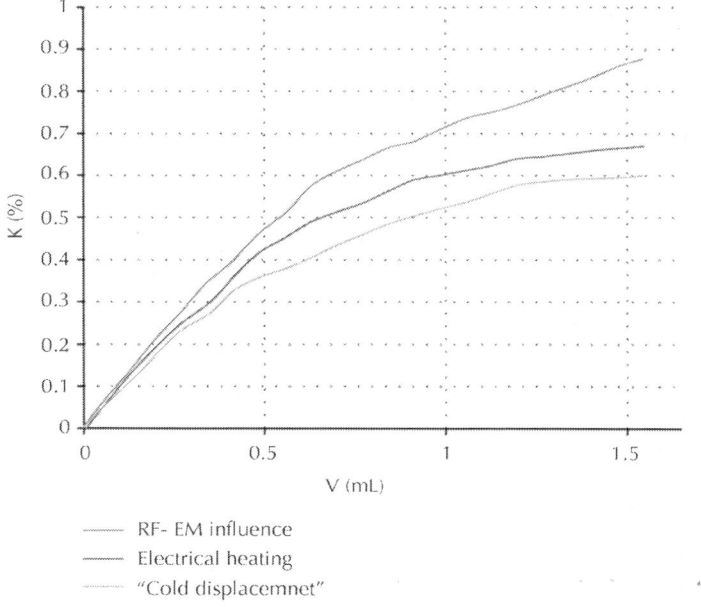

RF- EM influence
Electrical heating
"Cold displacemnet"

Figure 1: Dependence of oil recovery (K) on the relative volume of solvent injected. Adapted from—[6].

TECHNOLOGIES OF IN SITU ELEC-TROMAGNETIC HEATING OF HEAVY OIL AND BITUMEN

The first production method applying microwave heating to well production was patented in 1956 [14]. Electromagnetic waves were transferred to the well bottom from the surface through a coaxial system of internal and external pipes (tubing and casing). Interaction of electromagnetic waves with the formation causes the emergence of distributed volumetric heat sources and reduces the viscosity of the reservoir fluid. In 1965, Haagensen [13] described a device for generating high-frequency electromagnetic waves at the mouth of the well and a method of delivering electromagnetic energy through coaxial lines and waveguides to the bottom hole. In 1987, Wilson [16] described a similar device in his work, with some modifications of the radiating element of EM waves.

A huge drawback to the methods described by Haagensen and Wilson [13, 16] is the shallow penetration of electromagnetic waves, and, hence, a low sweep efficiency of heating. When the method described by Ritchey [14] was implemented, there were large losses of electromagnetic energy. Due to the finite conductivity of tubing, they are heated and electromagnetic energy is dissipated in rocks surrounding the well, resulting in large wellbore heat losses, especially if a permafrost layer is present.

Sayakhov et al. [8] proposed a method of recovery that included creating a combustion front with a simultaneous electromagnetic current influence. The environment is heated when exposed to an electromagnetic field, which decreases the viscosity and increases the mobility of crude oil. It is assumed in this method that the EM field continues influencing the reservoir after the combustion is initiated.

Review of Electromagnetic Heating Field Studies

Electromagnetic heating field trials have been carried out in Russia (Bashkortostan and Tatarstan) [2, 9, 10], the United States (California and Utah) [11, 17], and in Canada (Alberta and Saskatchewan) [18–21].

Russia

In Russia, field tests of radio frequency electromagnetic heating of the near-wellbore zone were first launched in 1969 at Well 40/19 in the Ishimbayskoye Oil Field in Bashkortostan and continued in the Yultimirovskoye Bitumen Field in Tatarstan according to Sayakhov et al. [2, 9, 10]. Characteristics of Well 40/19 from the Ishimbayskoye Oil Field are represented in Table 1.

Table 1: Characteristics of Well 40/19 of the Ishimbayskoye Oil Field

Parameter	Value
Depth, m	830
Casing diameter, in.	6
Tubing diameter, in.	2
Flow rate, ton/day	3
Well temperature, K	287–289
Paraffin content, %	2.3
Resins content, %	11
Density, kg/m^3	890
Viscosity at 20°C, m^2/c	$20 * 10^{-6}$

The source of high-frequency electromagnetic energy was a generator providing an optimum oscillation output power of 63 kW at a frequency of 13.56 MHz. Standard RF coaxial cable was used to supply high-frequency electromagnetic energy from the generator to the well. Temperature measurements were carried out using a thermograph at the bottom hole. Temperature was recorded continuously at a fixed depth of 650–655 m (in the open hole, where the radiating element was working) while the generator was operating (Table 2).

Table 2: Dynamics of temperature growth

Heating time, days	Temperature increase at the bottom hole, K
0.5	283
1	290
2	300
3	306
4	311
5	313

Yultimirovskoye Bitumen Field

RF electromagnetic heating was conducted in the Yultimirovskoye Bitumen Field by Sayakhov in 1980 [2] (Table 3). Two wells spaced

5 meters apart were studied, Well 150 and Well 1.

Table 3: Characteristics of the Yultimirovskoye reservoir

Parameter	Value
Porosity, %	25
Bitumen saturation, %	3.6
Permeability, micm2	0–0.183

Electromagnetic heating of the bitumen reservoir was conducted in several stages at different conditions. Initially, the RF-EM installation was set to about of 20 kW power. After 36.5 hours, the temperature at the bottom of Well 150 has increased from 282 to 389 K. No temperature change was observed in Well 1. In the next phase, RF-EM installation was reset to approximately 30 kW of power. As a result, the temperature in Well 150 reached 423 K after six hours. It should be noted that the growth rate of temperature increased. In the third phase, the RF-EM installation was set to a maximum of 60 kW, which caused the heating intensity in Well 150 to increase greatly. After 5.5 hours, the temperature in Well 150 increased from 417 to 463 K. Next, the RF-EM unit was turned off for 2 hours, resulting in a temperature drop to 423 K.

During the next 32 hours, the RF-EM installation operated at maximum output, causing the heating of the bottom hole of Well 150 up to 583 K and the bottom hole of Well 1 up to 318 K. In Sayakhov's experiment [2], the fluoroplastic collars that centered the tubing in Well 150 melted (their maximum operating temperature was 573 K). As a result, a short circuit between the casing and tubing occurred and the RF-EM unit broke and was disabled. Before this disruption, the RF-EM installation worked steadily throughout the field experiment.

Deep heat penetration (up to 5 m in the reservoir) was demonstrated by temperature measurements done during the cooling of Well 150's drain zone after the exposure. The temperature decreased from 373 K to 343 K in three days. After a few days of electromagnetic heating, wellbore heat distribution revealed low heat losses to both the overburden and underburden.

The United States

Bakersfield, California, United States

In 1992, Kasevich et al. [17] conducted field tests of RF-EM heating in the United States in the Bakersfield, California, field. The goal was to prove the concept that controlled RF-EM radiation could be used as a thermal EOR method. The production of reservoir fluids was not measured because it was a quality, not quantity, study designed to gain a better understanding of underground processes.

A high-frequency electromagnetic wave generator with a capacity of 25 kW and frequency of 13.56 MHz was used to heat the reservoir at Well 100D. Heat penetration was determined by temperature measurements in the surrounding observation wells T10, T20, and T30, which were located 3, 6, and 9 meters from Well 100D. Kasevich et al. [17] also proved that the RF producer used could efficiently focus its radiation pattern into the desired region by measuring return loss and electromagnetic radiation. In well T10, situated 3 meters from Well 100D, the medium temperature was increased from 293 K to 393 K over 20 hours of EM heating.

Avintaquin Canyon and Asphalt Ridge (Utah)

In 1980, one of the most detailed studies on electromagnetic-heat-based oil recovery was done at the Illinois Institute of Technology Research Institute (IITRI) by Bridges et al. [11]. They carried out extensive research work on the use of the different types of electromagnetic heating for different types of deposits, oil shale, and tar sand.

Bridges et al. [11] tested their IITRI technique of RF electromagnetic heating with two field experiments in Avintaquin Canyon, Utah, USA. Shale that was 6 meters thick was found in outcrops convenient for relatively cheap horizontal experiments. So arrays of holes were drilled and electrodes were inserted to a depth of 1 meter. These tests allowed the researchers to gain experience and to prove it was possible to achieve in situ pyrolysis of oil shale, thus increasing its thermal maturity. The power applied to the formation ranged from 5 kW to 20 kW, with a frequency of 13.56 MHz. As a result of EM heating, temperatures rose to 673 K and 20–30 % of the oil content was collected. However, it should be noted that the amount of produced oil was badly affected by the presence of cracks which allowed light hydrocarbons to escape (evaporate).

In 1981, Bridges et al. [11] conducted field tests on tar sand at Asphalt Ridge, Utah, USA. The first experiment tested the gravity drive bitumen recovery process, designed to prove EM heating concepts and improve equipment design. This experiment used vertical electrode placement and a mined collection chamber and tunnel. It was equipped with a 200 kW radio transmitter and heated 25 m^3 of tar sand. In the first experiment, the roof of the mined chamber was not supported well enough, which resulted in early termination of the experiment. Therefore, the situation had to be fixed by constructing a concrete arch for the second pilot test. The heating power used on tar sand varied from 40 kW to 75 kW with a frequency of 13.56 MHz.

The second test quantified the results of heating over a longer period and at higher temperatures. In this experiment, temperatures exceeded 473 K, and 30 to 35% recovery was achieved in just 20 days. This was encouraging because continuation of heating could have resulted in even higher recovery. The power loss was minimal in all the experiments, which proved the efficiency of heating.

CANADA

The Wildmere Field, Alberta, Canada

According to Spencer [18] commercial EM heating was first introduced in the field at Wildmere, Alberta, Canada (Table 4). The first well was drilled in January 1986 and began producing oil in March of the same year. Before EM heating began in May, the well was producing about 0.95 tonnes/day. After EM heating commenced, production rates increased and soon settled at the level of 3.18 tonnes/day until November 1986, when the well was closed due to technical reasons. Another well in this field increased in production from 1.59 tonnes/day to an average of 4.77 tonnes/day, with the maximum flow rate reaching 9.54 tonnes/day.

Table 4: Characteristics of the reservoir in the Wildmere field, Alberta, Canada

Parameter	Value
Net thickness, m	1
Depth, m	600
Density, kg/m3	987
Oil viscosity at 20°C, Pa·s	20

The Lloydminster Heavy Oil Area, Saskatchewan, Canada

In 1988-1989, two electromagnetic stimulation projects were conducted by Davidson [22] in the Lloydminster heavy oil area in Saskatchewan, Canada. Unfortunately the economic potential of the process could not be evaluated from either pilot test, because long-term heating could not be achieved due to equipment failure (casing insulation) and special reservoir conditions. However, the technical results looked promising.

The first pilot well was in Northminster (Saskatchewan, Canada) and produced 11.4 API oil from the Sparky formation. The power was applied to the well in a pulsating manner with a baseline of 20 kW with four-hour spikes of 30 kW (2 daily), in order to reduce the risk of significant damage to the insulation. Later, the power was increased to a 25 kW baseline with 35 kW pulses and finally to a 30 kW baseline with 50 kW peaks. At that point the insulation failed, and the power rates came down to 28 kW. Power rates were later leveled at 47 kW and stayed that way until terminated.

As can be seen from Table 5, water cut and production both reacted positively to electromagnetic stimulation. However, it should be noted that some portion of the increased production is related to the increase in pump speed. Water cut drop can be related directly to the EM effect and improvement in oil mobility. Once the heating has been terminated, technical parameters return rapidly to their initial states.

Table 5: Oil field performance for the Northminster pilot: primary and achieved by EM heating

Parameter	Primary	EM heating
Production rate, m³/day	10–12	20
Water cut, %	15–20	10–12
Productivity index, bbl/psi	0.33	0.42
Stimulation ratio		1.27

The second pilot well was situated in Lashburn (Saskatchewan, Canada) and produced very viscous 11.4 API crude oil from the Sparky formation. During the reservoir heating phase, the power ranged from 13 to 18 kW. This well has high sand cuts and before electrical power was applied its production was not stable and regularly had to be stimulated by flushing the wellbore. The peak production reached 5.0 m³/day and had begun to decline before electromagnetic heating was applied. Electromagnetic heating reduced the water cut, but the well was still prone to high water cuts after shut-in periods. Oil production also increased when electromagnetic forces were applied, until it reached 9 m³/day. During the initial heating phase of the reservoir, the temperature in the bottom hole increased steadily from 295 K to 309 K, but temperatures began dropping immediately after the power was turned off or when power delivery systems failed.

ESEIEH (Alberta, Canada)

Enhanced Solvent Extraction Incorporating Electromagnetic Heating technology (ESEIEH) has been patented and is currently undergoing tests, as reported by Rassenfoss [20]. The ESEIEH consortium is relying on three oil company partners to help with this testing: Laricina Energy, Nexen, and Suncor Energy. The pilot project is planned to take three years and currently is in the first stage. Field application is expected to start later in 2013. The ESEIEH method combines the familiar horizontal well pair design commonly used in the Canadian oil sands, coupled with heating using RF-EM waves and solvents, such as butane or propane. The company aims to heat the reservoir by running an antenna underground that emits enough energy to raise the temperature to 50°C (120°F).

CONCLUSIONS

A review of electromagnetic heating for enhanced oil recovery was presented in this paper. A number of studies show that electromagnetic heating is a promising method of enhanced oil recovery. However, the studies to date are limited, and only a few field trials have been reported. Most of the current research is based on laboratory experiments or numerical models. It should be noted that this paper did not cover the computer simulations carried out to research the effectiveness of EM heating.

Better understanding of the in situ electromagnetic process is essential and can be achieved by combining laboratory, numerical, and field-scale tests. At the moment it is not possible to assess the efficiency of EM heating or the opportunities for economic applications of it alone or in combination with traditional methods; therefore, more global studies should be conducted.

Even though sustainability of this technology has not yet been completely evaluated, the method definitely should not be overlooked by the industry because of its enormous potential. Attempts should be made to develop viable screening criteria for possible production of heavy oil, oil shale, and tar sand deposits.

REFERENCES

1.   S. Chistyakov, F. Sayakhov, and G. Balabyan, "Experimental study of formations dielectric properties under the influence of high-frequency electromagnetic fields," in University Investigations: Geology and Exploration, pp. 153–156, 1971.

2.   F. Sayakhov, "Particular properties of filtration and fluid flow under the influence of high-frequency electromagnetic field," in Joint University Scientific Book, pp. 108–120, 1980.

3.   B. Savinikh, V. Dyakonov, and A. Usmanov, "The influence of alternating electric currents on the thermal conductivity of dielectric fluids," Journal of Engineering Physics and Thermophysics, no. 2, pp. 269–276, 1981 (Russian).

4. A. Davletbaev and L. Kovaleva, "Combined RF EM/solvent treatment technique: heavy/extra-heavy oil production model case study," in Proceedings of the 10th Annual International Conference Petroleum Phase Behavior and Fouling, Rio de Janeiro, Brazil, 2009.

5. M. A. Fatikhov, "Experimental study of bitumen initial pressure gradient in the electromagnetic field," University Investigations: Oil and Gas, no. 5, pp. 93–94, 1990 (Russian).

6. L. Kovaleva, A. Davletbaev, T. Babadagli, and Z. Stepanova, "Effects of electrical and radio-frequency electromagnetic heating on the mass-transfer process during miscible injection for heavy-oil recovery," Energy and Fuels, vol. 25, no. 2, pp. 482–486, 2011.

7. G. Malofeev, O. Mirsaetov, and I. Cholovskaya, "Injection of hot fluids for enhanced oil recovery and well stimulation," in Regular and Chaotic Dynamics, Institute of Computer Science, Russialgevsk, Russia, 2008.

8. F. Sayakhov, R. Bulgakov, V. Dyblenko, B. Deshura, and M. Bykov, "About HF heating of bitumen reservoirs," Petroleum Engineering, no. 1, pp. 5–8, 1980 (Russian).

9. F. L. Sayakhov, L. A. Kovaleva, M. A. Fatikhov, and G. A. Khalikov, "Method of thermal effect on oil-bearing formation," SU Patent 1723314, 1992.

10. F. Sayakhov, I. Habibullin, M. Yagudin, and M. Fatyhov, "Technique and technology of thermal well stimulation on the basis electro-thermo-chemical and electromagnetic effects," University Investigations: Oil and Gas, no. 2, pp. 33–42, 1992 (Russian).

11. J. E. Bridges, J. J. Krstansky, A. Taflove, and G. C. Sresty, "The IITRI in situ RF fuel recovery process,"Journal of Microwave Power, vol. 18, no. 1, pp. 3–14, 1983. View at Scopus

12. J. Bridges, "Method for in-situ heat processing of hydrocarbonaceous formation," US Patent 4140180, 1979.

13. A. D. Haagensen, "Oil well microwave tools," Patent USA 3170119, 1965.

14. H. W. Ritchey, "Radiation Heating System, US Patent," Tech. Rep. 2757738, 1956.

15. G. C. Sresty, R. H. Snow, and J. E. Bridges, "Recovery of liquid hydrocarbons from oil shale by electromagnetic heating in-situ," US Patent 4485869, 1984.

16. R. Wilson, "Well production method using microwave heating," US Patent 4485868, 1987.

17. R. S. Kasevich, S. L. Price, D. L. Faust, and M. F. Fontaine, "Pilot testing of a radio frequency heating system for enhanced oil recovery from diatomaceous earth," in Proceedings of the SPE Annual Technical Conference & Exhibition, pp. 105–113, New Orleans, La, USA, September 1994.

18. H. L. Spencer, "Electromagnetic Oil Recovery, Ltd," Calgary, Canada, 1987.

19. F. E. Vermeulen and F. S. Chute, "Electromagnetic techniques in the in-situ recovery of heavy oils,"Journal of Microwave Power, vol. 18, no. 1, pp. 15–29, 1983.

20. S. Rassenfoss, "Seeking more oil, fewer emissions," Journal of Petroleum Technology, vol. 64, no. 9, pp. 34–38, 2012.

21. B. C. W. Mcgee and F. E. Vermeulen, "The mechanisms of electrical heating for the recovery of bitumen from oil sands," Journal of Canadian Petroleum Technology, vol. 46, no. 1, pp. 28–34, 2007.

22. R. J. Davidson, "Electromagnetic stimulation of Lloydminster heavy oil reservoirs: field test results,"Journal of Canadian Petroleum Technology, vol. 34, no. 4, pp. 15–24, 1995.

23. A. Chakma and K. N. Jha, "Heavy-oil recovery from thin pay zones by electromagnetic heating, paper SPE 24817," in Proceedings of the Annual Technical Conference and Exhibition, Society of Petroleum Engineers, Washington, DC, USA, October 1992.

24. B. Hascakir, C. Acar, Schlumberger, B. Demiral, and S. Akin, "Microwave assisted gravity drainage of heavy oils," in Proceedings of the International Petroleum Technology Conference (IPTC '08), pp. 1908–1916, Kuala Lumpur, Malaysia, December 2008.

25. B. Hascakir, T. Babadagli, and S. Akin, "Experimental and numerical modeling of heavy-oil recovery by electrical heating, paper SPE 117669," in Proceedings of the International Thermal

Operations and Heavy Oil Symposium (ITOHOS '08), p. 14, Society of Petroleum Engineers, Alberta, Canada, October 2008.

26. M. Koolman, N. Huber, D. Diehl, and B. Wacker, "Electromagnetic heating method to improve steam assisted gravity drainage, paper 1177481," in Proceedings of the International Thermal Operations and Heavy Oil Symposium (ITOHOS '08), pp. 327–338, Society of Petroleum Engineers, Alberta, Canada, October 2008.

27. K. A. Jha, N. Joshi, and A. Singh, "Applicability and assessment of micro-wave assisted gravity drainage (MWAGD) applications in Mehsana heavy oil field, paper SPE 14591," in Proceedings of the SPE Heavy Oil Conference and Exhibition, Society of Petroleum Engineers, Kuwait City, Kuwait, December 2011.

28. J. R. Kershaw, G. Barrass, and D. Gray, "Chemical nature of coal hydrogenation oils part I. The effect of catalysts," Fuel Processing Technology, vol. 3, no. 2, pp. 115–129, 1980.

29. S. Odenbach, "Ferrofluids—magnetically controlled suspensions," Colloids and Surfaces A, vol. 217, no. 1–3, pp. 171–178, 2003.

30. C. Ovalles, A. Fonseca, A. Lara et al., "Opportunities of downhole dielectric heating in Venezuela: three case studies involving medium, heavy and extra-heavy crude oil reservoirs, paper SPE 78980," in Proceedings of the International Thermal Operations and Heavy Oil Symposium and International Horizontal Well Technology Conference, Alberta, Canada, November 2002.

31. M. A. Ayrapetyan, "About oil fields development prospects by high-frequency currents electrical fields," in Materials of KSSR Institute of Oil, pp. 38–52, 1958.

32. M. A. Ayrapetyan, V. S. Velikanov, and E. Ya. Magnikov, "Reservoir high-frequency heating investigations," in Materials of KSSR Institute of Oil, pp. 113–124, 1959.

33. M. A. Carrizales, L. W. Lake, and R. T. Johns, "Production improvement of heavy-oil recovery by using electromagnetic heating, paper SPE 115723," in Proceedings of the SPE Annual Technical Conference and Exhibition (ATCE '08), Denver, Colo, USA, September 2008.

34. A. D. Hiebert, F. E. Vermeulen, F. S. Chute, and C. E. Capjack, "Numerical simulation results for the electrical heating of

Athabasca oil-sand formations," SPE Reservoir Engineering, vol. 1, no. 1, pp. 76–84, 1986.

35. J. Burge, P. Surio, and M. Combarnu, Thermal Methods of Enhanced Oil Recovery, Nedra Publishing, Moscow, Russia, 1988.

13

Microemulsions: A Novel Approach to Enhanced Oil Recovery: A Review

Achinta Bera[1] and Ajay Mandal[2]

[1]Department of Civil and Environmental Engineering, School of Mining and Petroleum Engineering, University of Alberta, Edmonton, AB T6G 2W2, Canada

[2]Department of Petroleum Engineering, Indian School of Mines, Dhanbad 826004, India

ABSTRACT

The trend of growing interest in alternative source of energy focuses on renewable products worldwide. However, the situation of petroleum industries in many countries needs much concern in improving the oil recovery technique. Chemical method, especially microemulsion flooding, plays an important role in enhanced oil recovery technique due to its ability to reduce interfacial tension between oil and water to a large extent as well as alter wettability of reservoir rocks. Surfactant-based chemical systems have been reported in many academic studies and their technological implementations are potential candidates in enhanced oil recovery activities. This paper reviews the role of different types of surfactants in enhanced oil recovery, structure of microemulsion, phase behavior of oil–brine–surfactant/cosurfactant systems with variation of different parameters such as salinity, temperature, pressure and physicochemical properties of microemulsions including solubilization capacity, interfacial tension, viscosity and density under reservoir conditions. The enhanced oil productivity by microemulsion flooding with different surfactant/cosurfactant systems has also been discussed in this paper. This review introduces a new opening in enhanced oil recovery by microemulsion flooding with some new aspects.

INTRODUCTION

The energy demand will be met by a global energy mix that is undergoing a transition from the current dominance of fossil fuels to a more balanced distribution of energy sources. New discoveries of conventional oil fields are declining, while demand for oil is increasing day by day, particularly in the developed and developing countries. After conventional waterflood processes, the residual oil in the reservoir remains as a discontinuous phase in the form of oil drops trapped by capillary forces and is likely to be around 70 % of the original oil in place (OOIP) (Dosher and Wise 1976). However, technically it is possible to improve this recovery efficiency by applying enhanced oil recovery (EOR) processes. Microemulsion is an efficient tool in EOR techniques because of its high level of extraction efficiency by reducing oil–water interfacial tension (Santanna et al. 2009; Bera et al. 2014a). Microemulsions are transparent and translucent homogeneous mixtures of hydrocarbons and water with large amounts of surfactants (Schulman et al. 1959; Stoeckenius et al. 1960). Alkanols (medium chain alcohols such as propanol, butanol, isoamyl alcohol, pentanol, hexanol, etc.) are generally used as cosurfactants for the preparation of microemulsions (Barakat et al. 1983; Lalanne-Cassou et al. 1983). The solubility of alkanol in water depends on the alcohol chain length. Short-chain alcohols such as methanol and ethanol are able to undergo a miscibility process with water. On the other hand, medium or long-chain alcohols (from propanol to higher alcohol) show very low solubility in water. In recent years, microemulsion flooding has become immensely important in the petroleum industries for the EOR technique (Santanna et al. 2009; Southwick et al. 2010; Kumar and Mohanty 2010; Flaaten et al. 2010; Elraies et al. 2010; Jeirani et al. 2013a, b; Bera et al. 2014b). Various research projects are involved in this field aiming to improve the petroleum oil recovery from natural oil reservoirs. The oil recovery process can be divided into mainly three stages such as primary recovery, secondary recovery and tertiary recovery (Gurgel et al. 2008). In primary oil recovery, oil is recovered due to pressure maintenance, in which oil is forced out through the production well by natural forces and reservoir gravity. The intrinsic or natural capacity of oil fields for producing oil is, however, promoted via primary recovery techniques. But physical constraints such as reduced well pressure and extensive oil trapping lessen oil production which eventually ceased at one stage and caused changes in the composition of crude oil affecting the reservoir wettability (Yangming et al. 2003). Chilingar and Yen (1983) thoroughly investigated different reservoir cores such as limestone, dolomite limestone, calcite dolomite and dolomite and concluded that 15 % were strongly oil wet, 65 % were oil wet, 12 % moderate oil wet and 8 % water wet. Therefore, in this point of view, it is an important issue regarding wettability for further desired oil recovery. When the required pressure is not available to expel the oil, water is injected to create pressure to recover the oil.

This is generally called secondary oil recovery or water flooding. The primary and secondary oil recovery process can recover nearly 30–35 % of OOIP. To recover the remaining oil, a tertiary recovery process is used. This tertiary oil recovery is also known as enhanced oil recovery. Tertiary EOR technique can be divided into mainly three categories, viz. chemical flooding, thermal process and gas injection. Alkali flooding, polymer flooding and micellar–polymer flooding are examples of chemical flooding. On the other hand, in situ combustion, steam injection and wet combustion methods are grouped into thermal processes (Leung et al. 1985; Sharma and Shah 1985; Auvray et al. 1984; Scriven 1976). Over the years, a number of innovative EOR processes such as microbial enhanced oil recovery (MEOR) and ultrasonic vibration methods have been introduced noticeably.

Another important technique is foam flooding. Foam is used during gas flooding such as with steam, CO_2 and miscible gas for mobility control. Sometimes, steam foams are used extensively to improve vertical and areal sweep efficiency and to reduce steam channeling in a shallow heavy oil reservoir. The steam foam may consist of surfactant with or without noncondensable gas. Steam foams have been used in conjunction with both continuous and cyclic steam injection. The classifications of EOR methods are shown in Fig. 1.

Capillary forces are also the important parameters for recovery of residual oil. These capillary forces are normally quantified by Young–Laplace equations in interfacial sciences (Schramm et al. 2003). The phase behavior of surfactant/cosurfactant–brine–oil system is the key factor in interpreting the performance of oil recovery by the microemulsion flooding process. Due to the well-established relationship between the microemulsion phase behavior and interfacial tension (IFT), it is common in the industry to screen surfactants and their formulations for low IFT through oil–water phase behavior tests (Shah 1981; Levitt et al. 2006; Engelskirchen et al. 2007; Kayalia et al. 2010).

The development of microemulsions for specific oilfield applications requires a systematic study of phase behavior as an important tool to select a treatment composition that satisfies specific parameters defined by the application. To identify microemulsion phase boundary, it is very common to study phase behavior in the laboratory. The formation and stability of microemulsion systems are driven by very low water–oil interfacial tension to compensate the large increase in the dispersion entropy (Bumajdad and Eastoe 2004). In the past several years, it has been shown that the phase behavior of surfactant/cosurfactant–oil–brine/water system is of the intense importance in the interpretation and forecasting the scopes of applications in the field of EOR techniques (Shah and Schechter 1977). At present, it is common that formulation of surfactant/cosurfactant–brine– oil systems that exhibit desirable phase behavior is an important stage in optimizing the performance of microemulsion systems for EOR methods (Healy and Reed 1974; Healy et al. 1975; Bera et al. 2012a). In surfactant/ cosurfactant–oil–brine systems, microemulsion shows different phase behavior with variation of different parameters such as salinity, temperature and pressure. The commonly observed Winsor-type system (Winsor 1954; Abe et al. 1987; Nakamae et al. 1990; Bera et al. 2012b) indicates that the microemulsions can exist in equilibrium with excess oil, excess water or both. In a Winsor-type I system, lower phase microemulsion exists with excess oil, and in case of Winsor-type II, upper phase microemulsion exists with excess brine. In general, middle-phase microemulsion (Healy et al. 1976) (surfactant-rich middle phase) system is known as Winsor-type III microemulsion and has a bicontinuous structure made by an equal mixture of water-inoil and oil-in-water type of microemulsions (Schulman et al. 1959; Stoeckenius et al. 1960). The single-phase microemulsion region is called Winsor-type IV phase. To represent these four phases, another notation system, especially employed by Kahlweit et al. (1990), uses the symbols 2; 2; 3 and 1, respectively. The factors that affect the phase transition between different types of systems include the salinity, temperature, molecular structure and nature of the surfactant and cosurfactant and the nature of the oil and water–oil ratio (WOR) (Shah 1985).

In this review paper, the phase behavior of surfactant/ cosurfactant–oil–brine system and the factors that affect the phase behavior and interfacial tension of the systems have been discussed. This paper also reviews the role of interfacial tension in oil recovery and its relation with phase behavior. In this connection, other properties of microemulsions (such as solubilization capacity, density and viscosity) that are directly or indirectly related with oil recovery have been also discussed in this review paper.

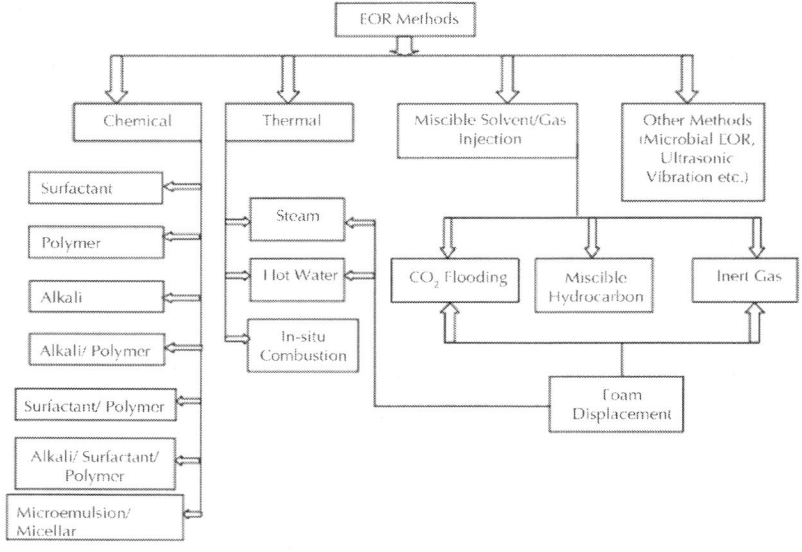

Figure 1: Simplified classification of EOR methods.

SURFACTANTS IN EOR TECHNIQUES

The choice of surfactants for EOR is another factor for efficient extraction of trapped oil from oil reservoirs. Surfactant solutions for use in EOR can have high (2.0–10.0 wt%) or low (0.1–0.2 wt%) surfactant concentration. To lower the interfacial tension up to an ultralow value, low surfactant concentration system was used where the aqueous phase of the surfactant solution is about the apparent critical micelle concentration (CMC). In case of high surfactant concentration systems, a middle-phase microemulsion is formed that is in equilibrium with excess oil and excess brine. The basic components of this microemulsion are surfactant, water, oil, alcohol and salts (generally, NaCl is used). High surfactant concentrations in the injected plug result in a relatively small pore volume (about 3–20 %) compared to micellar solutions (15–60 %).

Different surfactants were used to verify their activities in EOR techniques by laboratory experiments. Screening of surfactants for

EOR in laboratory is done based on the phase behavior experiments. Screening of petroleum sulfonates, which are generally known as the most available, commercial types of surfactants and manipulation of their combination in a chemical slug have been always of particular interest to researchers. In 1984, Bostich et al. (1984) invented an aqueous petroleum sulfonate mixture, which contained, at least, two different petroleum sulfonates. The implemented sulfonates had an average molecular weight within the range of about 300–450 g/mol. The hydrocarbon portion of the sulfonate had an average aliphatic to aromatic portion ratio within the range of about 4–20 mol per mole. In microemulsion flooding, anionic surfactants were used extensively. Several surfactants have been used in laboratory experiments for their screening for use in large scale. Few of the surfactants and their activities are discussed in Table 1.

MICROEMULSIONS IN EOR

Compositions of Microemulsions

It is well known that microemulsions are generally composed of hydrocarbons, surfactants/cosurfactants and brine. Surfactants are considered to be the principal constituents of microemulsions and are adsorbed at the interface rather than in the bulk phase. Surfactants are classified into four groups based on the charge of the head group such as anionic, cationic, nonionic and zwitterionic. Anionic surfactants such as sodium dodecylsulfate (SDS) are negatively charged in nature, but a small cation sodium ion occupies the counterpart. Anionic surfactants are most widely used in oil recovery process. Their adsorption phenomena in sandstone and carbonate are different. Their adsorption in sandstone is relatively lower than that of carbonate (Zhang et al. 2006). The adsorption can be reduced by use of sodium carbonate with the anionic surfactants.

Table 1: Surfactants used in laboratory studies in enhanced oil recovery

Sl no.	Author(s)	Names of the surfactants and types	Brief descriptions
1	Holm (1971)	Sodium sulfonates (anionic)	It was concluded that a clear microemulsion can be formed over a wide range of temperature by using this surfactant, and the microemulsion showed efficient extraction efficiency at elevated temperature
2	Healy et al. (1975)	Monoethanol amine salt of alkylorthoxylene sulfonic acid (anionic)	They concluded that microemulsion flooding can be considered as a miscible process before breakdown of slug and then it will be immiscible. It was also stated that surfactant retention in porous media is a serious problem in microemulsion flooding
3	Healy and Reed (1977)	Nonyl surfactant (anionic)	It was reported that ultralow interfacial tension between microemulsion–oil systems increases the oil recovery efficiency

4	Glover et al. (1979)	Monoethanol amine salt of dodecylorthoxylene sulfonic acid (anionic)	Depending on the salinity surfactant retention varied from 0.16 to 0.65 mg/g and rock and phase behavior was affected by salinity; as salinity changes from low to high values, middle-phase to lower-phase inversion takes place
5	Meyers and Salter (1980)	A sulfonated petroleum derivative (anionic)	Dynamic adsorption study of the surfactant was reported in Berea sandstone at different oil/brine ratios
6	Willhite et al. (1980)	TRS 10–80 petroleum sulfonate (anionic)	In these microemulsion displacements a new milky microemulsion was produced by mixing of the injected one and it followed the oil bank. The efficiency of the microemulsion was quite high. The adsorption of the surfactant on rock surface was also significant
7	Verkruyse and Salter (1985)	Ethoxylated alcohols (nonionic)	The microemulsion made with this surfactant did not show the desired result of oil recovery, but reduced the interfacial tension and showed high solubilization capacity

8	Sayyouh et al. (1991)	WITCO TRS-18 petroleum sulfonate (anionic)	With this surfactant, flooding experiments were carried out and satisfactory results were recorded. Phase behavior was also investigated with this microemulsion
9	Maerker and Gale (1992)	A blend of two tridecyl alcohol sulfates (PL612 ? Pl613) (anionic)	This surfactant was used to prepare diesel oil microemulsion. This is a cost-effective formulation of microemulsion and gives good results in oil recovery experiments
10	Purwono and Murachman (2001)	Sodium ligno sulfonate (anionic)	The microemulsion formulated with the surfactant was useful, but the compositions are not available and so the use of this microemulsion is not always possible
11	Bouabboune et al. (2006)	Alkali surfactant NM (anioinic)	With this surfactant, a comparative study was conducted between surfactant and microemulsion flooding. The result showed a higher efficiency of microemulsion than the surfactant solution alone
12	Chai et al. (2007)	Sodium dodecyl sulfonate, sodium dodecyl sulfate and sodium dodecyl benzene sulfonate (all are anionic)	Phase behavior study of microemulsion was carried out with these surfactants. The formulated of the microemulsions showed different ε-β phase behaviors

13	Santanna et al. (2009)	Soap sodium soap and a synthetic surfactant (anionic)	The microemulsions are very much efficient to recover residual, and viscosity of the microemulsion also plays an important role here. The more viscous microemulsion led to more oil recovery than the other one with lower viscosity
14	Zhu et al. (2009)	Triton X100 (nonionic), cetyltrimethy-lammonium bromide (cationic)	Triton X-100 and its oligomer tyloxapol with cetyltrimethy-lammonium bromide induced by hydrolyzed polyacrylamide was used to produce ultralow interfacial tension between crude oil and aqueous phase (brine). This combined mixture is very much effective for lowering interfacial tension and applicable for additional oil recovery
15	Iglauer et al. (2010)	Di-tridecyl sulfosuccinic acid ester (Aerosol TR-70), coconut diethanolamide (CW-100), alkylpolyglycosides (Agrimul PG2062 and Agrimul PG2069), alkylpropoxy sulfate sodium salts (Alfoterra 23 and Alfoterra 38) (all are anionic)	The core flooding experiments give some encouraging results (15–75 %) of enhanced oil recovery

16	Wan et al. (2011)	Span-80 and Tween 80 (nonionic)	These surfactants were used to form a copolymer via an inverse microemulsion system. This copolymer was used as a drilling fluid
17	Trabelsi et al. (2011)	Triton X405 (nonionic), sodium dodecyl sulfate (anionic), sodium dodecylbenzene sulfonate (anionic)	Ultralow interfacial tensions were reached using these surfactants. It was also observed that addition of alkaline chemicals further reduces the interfacial tension between the oil and aqueous phases. Among these surfactants, SDBS performs better than the others
18	Qiao et al. (2012)	1,3,5-triazine surfactants (Nonionic)	The double long-chain 1,3,5-triazine surfactants are very much active in reducing interfacial tension up to ultralow value. Therefore these surfactants have significant importance in the EOR method
19	Mandal et al. (2012)	Tergitol 15-S-5, Tergitol 12-S-7, Tergitol 15-S-9, Tergitol 15-S-12 (nonionic)	These surfactants were used to prepare nanoemulsions. The prepared nanoemulsions show good efficiency in additional oil recovery from crude oil-saturated sand pack. The additional oil recovery was more than 30 % OOIP

| 20 | Gao and Sharma (2013) | Alkyl sulfate gemini surfactant (anionic) | These gemini surfactants are effective in reducing interfacial tension between oil and water. These surfactants exhibit extraordinary aqueous stability even in high salinity and hard brines. At low concentrations the surfactants also show ultralow interfacial tension. The results from this study showed the potential of utilizing these surfactants at low concentrations and in harsh reservoir conditions |
| 21 | Lu et al. (2014) | Guerbet alkoxy sulfate (GAS) surfactants, Guerbet alkoxy carboxylate surfactants (anionic) | The newly developed surfactants are efficient for enhanced oil recovery as they can reduce the interfacial tension significantly. The Guerbet alkoxy carboxylate surfactants are alternatives to the sulfate surfactants for circumstances where the reservoir temperature is high and alkali is not included in the formulation |

22	McLendon et al. (2014)	Branched nonylphenol ethoxylates (Huntsman SURFONICs N-120 and Huntsman SURFONICs N-150) and branched isotridecyl ethoxylate (Huntsman SURFONICs TDA-9) (anionic)	These surfactants are more soluble in CO_2. So these surfactants can form stable CO_2 foam easily which is a very appropriate candidate for EOR application for mobility control
23	Bai et al. (2014)	Sulfonate surfactant such as alkyl benzene, alkyl naphthalene, alkyl indane and alkyl phenanthrene (anionic)	All these surfactants are very active at low concentration. Synergistic effect was found when ethanolamine was used with the surfactants

Cationic surfactants are positively charged and they are easily adsorbed in anionic surfaces of clay and sand. Therefore, they are not extensively used in oil recovery process for sandstone reservoirs. In recent investigation, it has been proved that cationic surfactants like cetyltrimethylammonium bromide (CTAB) performed better than anionic surfactants in altering the wettability of the carbonate rock to a more water wet (Salehi et al. 2008). Nonionic surfactants do not form ionic bonds, but the ether groups of the nonionic surfactants can form hydrogen bonding with water so that nonionic surfactants show surface properties (Myers 2006). Therefore, these surfactants introduce their polarity from having an oxygen-rich portion of the molecule at one end and a large organic portion at the other. Nonionic microemulsions also produce ultralow interfacial tensions and show high solubilization parameter (Verkruyse and Salter 1985; Iglauer et al. 2009). Another type of surfactant, i.e., zwitterionic may contain both positive and negative charges. Recently, Wang et al. (2010) have applied such type of surfactant in surfactantpolymer flooding for enhanced oil recovery and have obtained good consistent results.

The unique properties of microemulsions differentiate them from ordinary emulsion. For formulation of microemulsions, different alkanes with carbon number from C6 to C18 are generally used. The physicochemical properties of microemulsions depend on the alkane carbon number, nature of cosurfactant and types of surfactants. For ionic microemulsion preparation, cosurfactants are added along with surfactant (Healy and Reed 1977; Willhite et al. 1980), but some other researchers did not consider cosurfactant as a main component (Nelson and Pope 1978; Osterloh and Jante 1992). Microemulsions actually contain a cosurfactant such as a medium chain alcohol (viz. propan-2-ol, butanol, isoamyl alcohol etc.) in combination with a primary surfactant (Barakat et al. 1983; Lalanne-Cassou et al. 1983). Owing to the diphilic nature of surfactants, they distribute their head and tail parts to the corresponding polar and nonpolar solvents forming a monolayer film between them. This fact decreases the interfacial tension to an ultralow value, which results in the solubilization of the otherwise immiscible components.

Cosurfactant, a second surfactant is generally added to the surfactant to prepare microemulsions due to its well-documented roles in various applications especially in EOR (Zhou and Rhue 2000; Richardson et al. 1997; Pattarino et al. 2000; Cavalli et al. 1996) such as to (1) prevent the formation of rigid structures such as gels, liquid crystals and precipitates; (2) alter the viscosity of the system; (3) reduce interfacial tension; (4) increase the fluidity of the interface and thereby increase the entropy of the system; (5) increase the mobility of the hydrocarbon tail and allow greater penetration of the oil into this region; (6) modify the hydrophilic–lipophilic balance (HLB) values of surfactants; and (7) influence the solubility properties of the aqueous and oleic phases due to its partitioning between the phases. Several attempts to substitute traditional cosurfactants with other components (Sagitani and Friberg 1980; Osborne et al. 1988; Comelles and Pascual 1997), for example nonionic surfactants, alkanoic acids, alkanediols, amines, aldehydes, ketones, butyl lactate and oleic acid, have been carried out from the viewpoint of suitable applications of microemulsions. The role of cosolvents and additives in the preparation of microemulsions has also been reported in literature (Kahlweit et al. 1985; Wormuth and Kaler 1987; Kim et al. 1988). Commonly, cationic, anionic and nonionic surfactants are used to form microemulsions. Physicochemical properties of the microemulsions are different for different surfactant systems. Brine is generally considered a pseudo component for microemulsion system taking in account water and salt as single phase.

Type and Structure of Microemulsion

Microemulsion structure has a key role in the different physicochemical parameters of the applied fields. The specific structures of the microemulsions have been extensively studied by many researchers (Azouz et al. 1992; Wadle et al. 1993; Maidment et al. 1997; Mo et al. 2000; Li et al. 2010). The three basic types of microemulsions are direct (oil dispersed in water, o/w), reversed (water dispersed in oil, w/o) and bicontinuous. Like multiple emulsion, sometimes multiple microemulsion are also possible. In this type, another layer is formed outside the o/w or w/o microemulsions. The schematic diagram of the basic three types of microemulsions is shown in Fig. 2. Microemulsion structure depends on salinity, water content, cosurfactant concentration and surfactant concentration. At higher water content, the microemulsion would be a water-external system with oil solubilized in the cores of the micelles. Although the mixtures remain single phase and thermodynamically stable, the microemulsion structure changes through a series of intermediate states (Bourrel and Schechter 1988). The structures of these intermediate states are not well known. However, the solutions are thermodynamically stable and isotropic. Salinity also can reverse the structure of the microemulsion. As salinity increases, the direct microemulsion changes to reverse microemulsion. At low salinity, the system remains in water-external phase, but with increasing salinity the system separates into an oil-external microemulsion.

Surfactant/Microemulsion Flooding

During the past 40 years, it has been reported that many surfactant formulations for EOR generally form multiphase microemulsions (Chiang and Shah 1980; Cayias et al. 1977; Wilson et al. 1976; Schwuger et al. 1975).

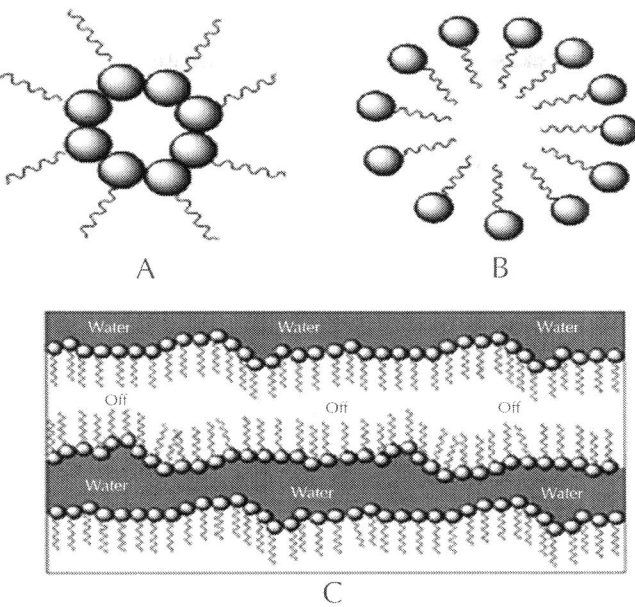

Figure 2: Microemulsion structures: a. reverse microemulsion, b. direct microemulsion, and c. bicontinuous microemulsion.

In chemical EOR methods, the process occurs with a certain degree of chemical interaction between the injected fluid and the reservoir fluid. This may be achieved by injecting polymer solutions, surfactant slugs, microemulsions or alkaline solutions. The main purpose of surfactant flooding is to reduce the interfacial tension between oil and water, thus increasing the displacement efficiency (Kwok et al. 1995). Surfactant solutions are not so efficient for better recovery of oil due to low viscosity compared to that of the oil. From this point of view, microemulsions are better replaceable injected fluids for their unique properties, which feature higher viscosity and induce low interfacial tension, increasing the oil extraction efficiency. Due to adsorption of surfactant molecules on the reservoir rock surfaces, low concentration of surfactant solutions are not allowable many times (Kassel 1989). Austad and Strand (1996) studied the microemulsion system and stated that very low interfacial tension may be reached with microemulsion systems. Under such circumstances, microemulsions flow more easily through the porous media, which enhances the oil extraction performance. For good efficiency of microemulsions and reasonable oil recovery efficiency, the surfactant must be chemically stable, reduce the interfacial tension between brine and crude oil and displace the oil without significant surfactant loss by adsorption on the reservoir rocks. Retention of surfactant is a most restrictive factor that affects the efficiency of the oil recovery process by microemulsion flooding (Glover et al. 1979). The microemulsion slug partitions into three phases such as a surfactant-rich middle-phase and surfactant-lean brine and oil phases (Healy and Reed 1974, 1977; Healy et al. 1976) in the intermediate salinity range. The surfactant-rich phase is the middle-phase microemulsion. In case of microemulsion flooding, a high concentration of surfactant must be used so as to form micelles that can solubilize or dissolve the reservoir oil. This phenomenon takes place via incorporation of small oil droplets in the micelle core, effectively causing miscibility in the system (Shindy et al. 1997).

Mechanism of Surfactant/Microemulsion Flooding

In microemulsion flooding, the reservoir is flooded with water containing a small percentage of surfactant and other additives such as hydrocarbon, medium-chain alcohol and brine. The surfactant plays a key role in forming the exact type of microemulsion that reduces the interfacial tension of the target oil (Robert Moene, Shell Global Solutions).

This is critical to both mobilize oil and enable it to escape from the rock. Generally speaking, whenever a waterflood has been successful, microemulsion flooding will be applicable, while in many cases where flooding has failed owing to its poor mobility relationships, microemulsion flooding can still be successful mainly due to the required mobility control. A schematic diagram of surfactant/microemulsion flooding is shown in Fig. 3. The microemulsion flooding process is implemented as tertiary displacement near the end of a water flood. Figure 3 shows a tertiary process where residual oil saturation exists. A specified volume of surfactant slug (0.5–1.0 PV) is injected.

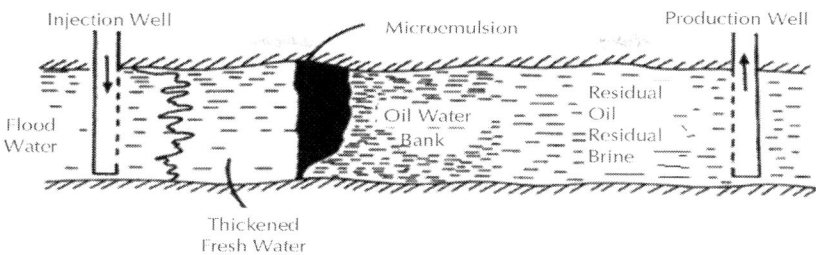

Figure 3: Two-dimensional schematic diagram of microemulsion flooding.

The micellar solution has a very low IFT with the residual oil and mobilizes the trapped oil, forming an oil bank ahead of the slug. The slug has a very low IFT with the brine and thus displaces brine as well as oil. Both oil and water flow in the oil bank. The thickened water is injected after the surfactant slug to drive the slug of microemulsion through the formation towards the production wells. A thickened water or mobility buffer slug consisting of a solution of polymer in water is used. The micellar solution must be designed in such a way that a favorable mobility ratio exists between the micellar slug and the oil bank. The viscosity of the micellar solution is adjusted to accomplish this. A polymer is often added to the micellar solution to increase its apparent viscosity. Thus, the process has the potential to increase both volumetric sweep efficiency and microscopic displacement efficiency. In some cases, a preflush is injected ahead of the micellar solution to adjust the brine salinity or pH. The preflush solution may contain a sacrificial adsorbent that will be adsorbed on the rock and occupy adsorption sites. The purpose is to reduce adsorption and loss of the surfactant contained in the micellar solution.

Phase Behavior of Surfactant/Cosurfactant–Brine–Oil System and Optimization Study

The phase behavior of surfactant–oil–brine system is the important key step in the laboratory to screen the proper surfactants for EOR. The microemulsion phase behavior changes from Winsor I to Winsor II through Winsor III with variation in salinity, temperature and pressure. Surfactant molecules in oil or in water form a variety of structures when structure-assisted parameters such as water content, surfactant concentration, cosurfactant type, cosurfactant concentration, pressure and/or temperature are varied. Figure 4 shows the schematic diagram of the Winsor phase behavior of microemulsion, and Fig. 5 depicts the corresponding pseudoternary phase diagram. The middle-phase microemulsion consists of solubilized oil, brine, surfactant and alcohol. The lower to middle to upper phase transition of the microemulsion phase can be obtained by varying the following factors: (1) increasing salinity, (2) increasing alcohol concentration (propanol, butanol, pentanol and hexanol), (3) decreasing oil chain length, (4) changing temperature, (5) increasing total surfactant concentration, (6) increasing surfactant solution/oil ratio, (7) increasing brine/oil ratio and (8) increasing molecular weight of the surfactant.

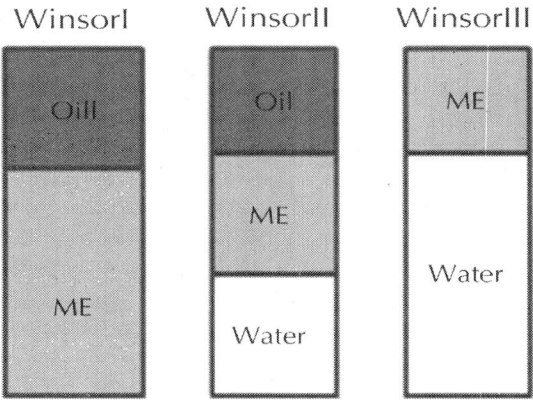

Figure 4: Schematic diagram of Winsor-type phase behavior of microemulsion (ME).

Hussain et al. (1997) studied the three-phase microemulsion systems as a function of temperature and pH. They showed that the presence of ethylene oxide (EO) moiety in the surfactant molecule made the surfactants less sensitive to salinity than an anionic surfactant. They also stated that the carboxylic ionic head group made the surfactant more stable to temperature than in simple EO nonionic surfactants. The phase behavior of pH-dependent ether carboxylic acid system depends on salinity in the same way as in ethoxylated sulfonates

(Qutubuddin et al. 1984). The middle-phase microemulsion is formed at low to high pH as a function of temperature at constant salinity. Increasing EO units in the surfactant molecules makes the surfactant molecule more hydrophilic, and hence high salinity and temperature are required to achieve the optimum region (Hussain et al. 1997). John and Rakshit (1994) studied the phase behavior and properties of cyclohexane/ SDS/propanol/water microemulsion system in the presence of NaCl. They reported that the one-phase microemulsion region disappears completely at higher NaCl concentration. Abe et al. (1992) investigated the phase behavior and physiochemical properties of sodium octyl sulfate/n-decane/1-hexanol/aqueous $AlCl_3$ middle-phase microemulsion.

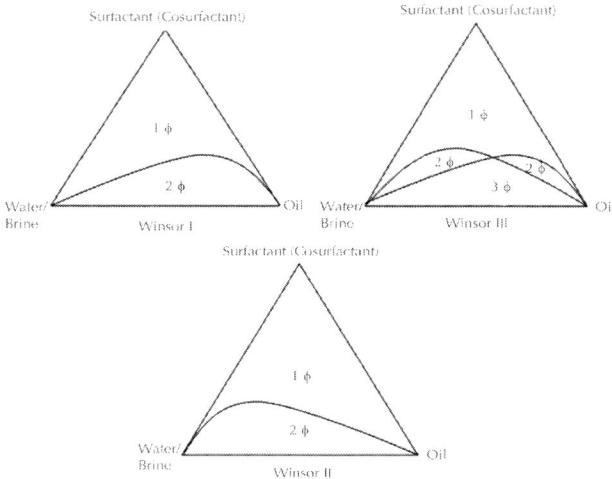

Figure 5: Pseudoternary phase diagram of oil–water (brine)–surfactant (co-surfactant) system (Winsor-type phase behavior of microemulsion where "/" indicates the number of phases).

They reported that at a particular salinity, a drastic change in the phase volume and physicochemical properties might be attributed to a phase inversion of $AlCl_3$ middle-phase microemulsion from oil-rich to water-rich continuous phase with increasing $AlCl_3$ concentration, which is quite a different behavior from that observed for monovalent and divalent salt systems. They also concluded that the nature of middle-phase microemulsion was very complicated and that its microstructure involves an intermicellar equilibrium incorporating various types of droplets, bicontinuous fluctuating and/or a rigid liquid crystal phase states, depending on the type of salt in the system. Bera et al. (2011) studied the interfacial tension and phase behavior of surfactant– brine–oil system for three nonionic surfactants. They reported that with increasing salinity, relative phase volume of middle-phase microemulsion would be increased due to increase conductivity of the solution and hydrophilicity of the molecules.

Solubilization Capacity of Microemulsion

High solubilization capacity of microemulsion depends on the formulation of microemulsion. Solubilization capacity of microemulsion is a function of surfactant category, oil type, cosurfactant nature, etc. Salts are also responsible for the variation of solubilization capacity of microemulsion. Microemulsions show a high level of solubilization capacity toward both oil and water simultaneously. This property makes them one of the most important tools in chemical EOR. In the presence of some additives, the water solubilization capacity of microemulsions is also influenced. Water solubilization of a microemulsion sometimes obeys the Bansal, Shah and O'Connell (BSO) equation. The equation suggests that maximum solubilization occurs if the summation of the carbon chain length of cosurfactant (l_{cs}) and oil (l_0) is equal to the carbon chain length of the surfactant (l_s), i.e., $l_{cs} + l_0 = l_s$. Recently in our few experimental studies, we also show the results of this rule (Bera et al. 2012c, 2014c).

Wei et al. (2011) reported that the water solubilization capacity increases initially with increasing ionic liquid (additive) concentration and reaches a maximum and then decreases with a further increase of ionic liquid concentration. They also reported the effect of the addition of salt to the microemulsion. The addition of salt diminishes the effective polar area of the surfactant by decreasing the thickness of the electrical double layer around the polar group. In this regard, packing parameter (P) is defined as v/al, where v is the effective volume of a surfactant molecule, a is the effective area of its polar head and l is the length of its hydrocarbon chain. As P increases, the water solubilization capacity of microemulsion decreases; because of the two counteracting factors, a maximum value of water solubilization capacity is observed. As the concentration of salt becomes higher, the latter effect plays a dominant role, which explains the further decrease in water solubilization capacity.

Interfacial Tension Reduction

Ultralow interfacial tension is required to recover the trapped oil with increasing capillary number. It is well known that ultralow interfacial tension plays an important role in oil recovery processes (Chiang and Shah 1980; Cayias et al. 1977; Wilson et al. 1976). The reasons of ultralow IFT's have been extensively investigated. The ultralow IFT are associated with phase behavior at plait point. At the plait point of liquid/liquid system, two phases become indistinguishable and IFT between the two equilibrium phases goes to zero. The microemulsion systems exhibit ultralow IFTs over wide ranges of salinities, surfactant concentrations and temperatures, suggesting that a critical phenomenon is involved. In 1977, Shah and Schechter (1977) demonstrated from their experimental results that there was direct correlation between interfacial tension and interfacial charge in various oil–water systems. Interfacial charge density is an important factor in lowering the IFT. Partition coefficient and IFT are strong functions of salinity. The minimum IFT occurs at the same salinity where the partition coefficient is observed to unity. Baviere (1976) proposed the same correlation between IFT and partition coefficient. Different factors (such as surfactant mixture ratio, salt concentration, temperature, pressure and oil types) influence the IFT. Different category surfactants have different activities to reduce IFT. Sometimes, mixed surfactants show extra ability to reduce IFT. Therefore, the mixing ratio plays an important role.

When interaction is not so strong, the two surface-active materials with equimolar concentration in the phase give the lowest IFT value among the different ratio mixtures (Rosen 1989; El-Batanoney et al. 1999). The relative solubilities of surfactant in oil and water vary significantly with change in the salinity of the aqueous phase. At low salt concentration, most of the surfactant molecules stay in the aqueous phase, while at high salt concentration, the surfactant molecules preferentially dissolve into the oil phase. An equal distribution of surfactant in both oil and water phase is observed at a particular salinity, called optimal salinity, which produces the lowest IFT (Sayyouh 1994). The solubility of surfactant molecule in aqueous medium is reduced by salt (Anderson et al. 1976). However, at a certain concentration of surfactant, the presence of NaCl salt (up to a certain concentration) may promote the surfactant migration toward the interfacial layer from the bulk phase, generating a substantial decrease in the IFT between oil water (Schechter and Wade 1976; Bera et al. 2014a, b, c). Therefore, the IFT decreases with increasing salinity up to a certain salt concentration and then increases. This salt concentration is generally known as optimal salinity.

Viscosity and Density of Microemulsion

The magnitudes of the viscosity and density of displacing fluid relative to the displaced fluid are important design variables that affect volumetric displacement efficiency. The tendency for gravity override and underride to occur is determined by relative densities of the displaced and displacing fluid. Areal and vertical sweep efficiencies are in large measure determined by the mobility ratio in the displacement process. Both viscosity and density are functions of microemulsion composition. Viscosity, in particular, can be varied over a wide range by proper adjustment of composition and/or by polymer addition. The viscosity of microemulsions depends on the structure of the microemulsion, i.e., whether it is water- or oilexternal. It is well known that at low water content, the system is oil-external, and at high water content, the system shows the reverse, i.e., water-external. The viscosity of the microemulsion increases as water content increases, creating swollen micelles. At the 50 % water content, the viscosity of the microemulsion increased to two orders of the initial value. At higher water content, after inversion to a water-external system, the viscosity decreases with further addition of water. In general, the viscosity of displacing slug has been modified by addition of a polymer, such as polyacrylamide or biopolymer. The viscosity of the microemulsion can be modified by adding a cosurfactant (medium chain alcohol) and/or polymer to the microemulsion.

APPLICATIONS OF MICROEMULSIONS IN ENHANCED OIL RECOVERY

The use of microemulsion is of high interest in many aspects of crude oil exploitation in EOR, especially due to the ultralow interfacial tension values attained between the contacting oil and water phases. Microemulsion flooding in EOR can be applied over a wide range of reservoir conditions due to its exclusive ultralow interfacial tension property (Pottmann 1974; Santanna et al. 2009; Jeirani et al. 2013a). In cases where the pressure exerted by water on the oil phase is not able to overcome capillary forces sufficiently, microemulsions are the key to extracting more than just a minor portion of crude oil. Properly balanced microemulsions are able to do so by drastically reducing the interfacial tension to the magnitude of 0.001 mNm^{-1}. This is also known as chemical flooding. Figure 6 shows the chemical enhanced oil recovery by micellar–polymer flooding.

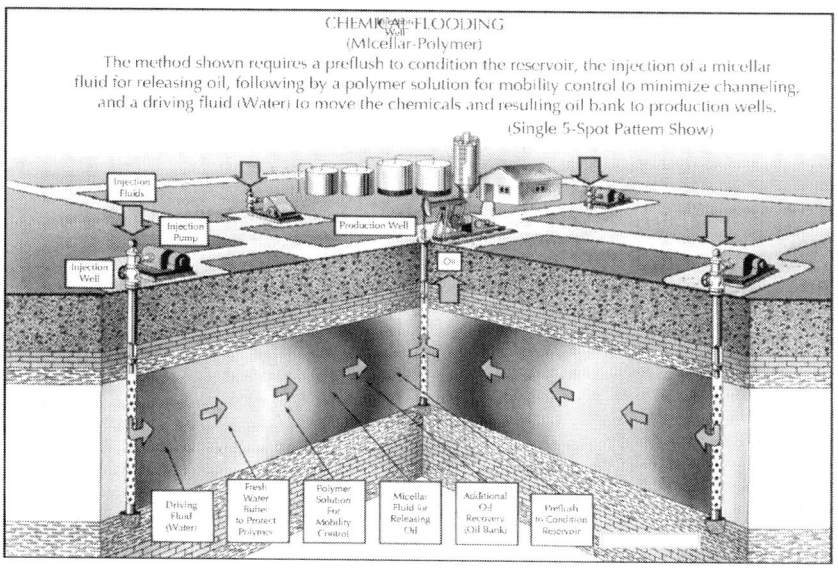

Figure 6: Micellar–polymer flooding technique for enhanced oil recovery.

Healy and Reed (1974) reported on some fundamentals of microemulsion flooding, especially viscosity, interfacial tension and salinity, relating the results of phase behavior of self-assembled systems to the Winsor's concepts. Austad et al. (1994a, b) discussed the physicochemical aspects involved in this method, particularly the interactions existing within specific polymer–surfactant and microemulsion systems applied in EOR. In microemulsion techniques, the oil reservoir is flooded with water containing a small percentage of surfactant and other additives. This solution reacts with natural acids in the trapped oil, making a microemulsion similar to soap lather. As a result, lowering of interfacial tension between oil and water has been found. This is critical to both mobilizing oil and enabling it to escape from the rock. Microemulsion is injected into reservoirs in EOR processes and lowers the IFT to mobilize the residual oil left trapped in the reservoirs after water flooding (Engelskirchen et al. 2007; Barnes et al. 2008; Santanna et al. 2009; Jeirani et al. 2013b). Tertiary oil recovery by means of microemulsions has been the main focus due to the ability to dissolve oil and water simultaneously in addition to attainment of very low interfacial tension of the system. Therefore, the design and analysis of chemical flooding processes for EOR depend on calculations of phase equilibria for these systems that are composed of water or brine, oil, surfactant and cosurfactant (usually a medium chain alcohol). Consequently, understanding the phase behavior of these systems is of fundamental importance to the development of any surfactant-based chemical flooding process. Microemulsions employed in EOR may be either oil-external (also called soluble oil) or water-external; mostly, they contain crude oil from the reservoir in which they are injected. The design of a microemulsion for a specific reservoir is basically a trial-and-error procedure; that is, the formulation of the microemulsion slug for a particular reservoir depends on the reservoir condition after the secondary recovery process and the properties of the microemulsion slug itself.

CONCLUSIONS

Microemulsions are very effective injecting fluids for extracting residual oil from reservoir in chemical EOR. Different chemical EOR techniques have been used for the last several decades in the oil recovery process. The drawbacks of the processes have forced to invent new techniques in EOR. Research in EOR is attributed to design and implementation of novel chemical methods. Mixtures, particularly of surface-active chemical substances, are incorporated in the injection formulations in this technology.

This aims at oil displacement that takes place due to attaining ultralow interfacial tensions and reduced fluid viscosity in oil reservoirs. Beside this, microemulsions show extraordinary water solubilization capacity which again makes them capable for excellent injecting fluids in chemical EOR techniques. Laboratory investigations are now successful for enhanced oil recovery from sand pack or original cores. But this microemulsion slug has not been injected into the field to test its efficiency. Only few pilot tests have been implemented till now and these have been successful. So it is recommended that microemulsion flooding has proper efficiency to extract oil from natural reservoirs and can be used for pilot tests.

Acknowledgments The authors gratefully acknowledge the financial assistance provided by the Council of Scientific and Industrial Research (CSIR) (Project: 22(0649)/013/EMR-II), New Delhi, and the Department of Petroleum Engineering, Indian School of Mines, Dhanbad, India. Thanks are also extended to all individuals associated with the project.

REFERENCES

1. Abe M, Schechter RS, Selliah RD, Sheikh B, Wade WH (1987) Phase behavior of branched tail ethoxylated carboxylate surfactant/electrolyte/alkane systems. J Dispers Sci Technol 8:157–172

2. Abe M, Yamazaki T, Ogino K (1992) Phase behavior and physicochemical properties of sodium octyl sulfate/n-dodecan/1-hexanol/aqueous AlCl3 middle-phase microemulsion. Langmuir 8:833–837

3. Anderson DR, Biduer MS, Dravis MT, Manning CD, Schevan LE (1976) Interfacial tension and phase behaviour in surfactant brine

oil systems. Presented at SPE Improved Oil Recovery Symposium, Tulsa, OK

4. Austad T, Strand S (1996) Chemical flooding of oil reservoirs-4. Effects of temperature and pressure on the middle phase solubilization parameters close to optimum flood conditions. Colloids Surf A 108:243–252

5. Austad T, Fjelde I, Veggeland K (1994a) Adsorption VI. Nonequilibrium adsorption of ethoxylated sulfonate onto reservoir cores in the presence of xanthan. J Pet Sci Eng 12:1–8

6. Austad T, Fjelde I, Veggeland K, Taugbol K (1994b) Physicochemical principles of low tension polymer flood. J Pet Sci Eng 10:255–269

7. Auvray L, Cotton JP, Ober R, Taupin C (1984) Structure and phase equilibria of microemulsions. J Phys 45:913–925

8. Azouz IB, Ober R, Nakache E, Williams CE (1992) A small angle X-ray scattering investigation of the structure of a ternary waterin-oil microemulsion. Colloids Surf A 69:87–97

9. Bai Y, Xiong C, Shang X, Xin Y (2014) Experimental study on ethanolamine/surfactant flooding for enhanced oil recovery. Energy Fuels 28:1829–1837

10. Barakat Y, Fortney LN, Schechter RS, Wade WH, Yiv S, Graciaa A (1983) Criteria for structuring surfactants to maximize solubilization of oil and water: II. Alkyl benzene sodium sulfonates. J Colloid Interface Sci 92:561–574

11. Barnes JR, Smith JP, Smith JR, Shpakoff PG, Raney KH, Puerto MC (2008) Phase behavior methods for the evaluation of surfactants for chemical flooding at higher temperature reservoir conditions. Paper SPE 113314, presented at the 2008 SPE/DOE Improved Oil Recovery Symposium, Tulsa, OK

12. Baviere M (1976) Phase diagram optimization in micellar systems. Presented at the 51st annual fall technical conference and exhibition of the society of petroleum engineers of AIME, New Orleans, USA

13. Bera A, Ojha K, Mandal A, Kumar T (2011) Interfacial tension and phase behavior of surfactant-brine–oil system. Colloids Surf A 383:114–119

14. Bera A, Ojha K, Mandal A, Kumar T (2012a) Phase behavior and physicochemical properties of (sodium dodecyl sulfate ? brine

? propan-1-ol ? heptane) microemulsions. J Chem Eng Data 57:1000–1006

15. Bera A, Kumar S, Mandal A (2012b) Temperature-dependent phase behavior, particle size and conductivity of middle-phase microemulsions stabilized by ethoxylated nonionic surfactants. J Chem Eng Data 57:3617–3623

16. Bera A, Ojha K, Kumar T, Mandal A (2012c) Water solubilization capacity, interfacial compositions and thermodynamic parameters of anionic and cationic microemulsions. Colloids Surf A 404:70–77

17. Bera A, Kumar T, Ojha K, Mandal A (2014a) Screening of microemulsion properties for application in enhanced oil recovery. Fuel 121:198–207

18. Bera A, Mandal A, Guha BB (2014b) Effect of synergism of surfactant and salt mixture on interfacial tension reduction between crude oil and water in enhanced oil recovery. J Chem Eng Data 59:89–96

19. Bera A, Kumar T, Mandal A (2014c) Physicochemical characterization of anionic and cationic microemulsions: water solubilization, particle size distribution, surface tension and structural parameters. J Chem Eng Data 59:2490–2498

20. Bostich JM, Hsieh WC, Koepke JW (1984) Process for enhanced oil recovery employing petroleum sulfonate blends. Patent number US 4446036 A

21. Bouabboune M, Hammouch N, Benhadid S (2006) Comparison between microemulsion and surfactant solution flooding efficiency for enhanced oil recovery in TinFouye oil Field. Paper 2006-058, presented at the Petroleum Society's 7th Canadian International Petroleum Conference and 57th Annual Technical Meeting, Calgary, Alberta, Canada

22. Bourrel M, Schechter RS (1988) Microemulsion and related systems. Marcel Dekker Inc., New York

23. Bumajdad A, Eastoe J (2004) Conductivity of mixed surfactant waterin-oil microemulsions. Phys Chem Chem Phys 6:1597–1602

24. Cavalli R, Marengo E, Caputo O, Ugazio E, Gasco MR (1996)

The effect of alcohols with different structures on the formation of warm o/w microemulsions. J Dispers Sci Technol 17:717–734

25. Cayias JL, Schechter RS, Wade WH (1977) The utilization of petroleum sulfonates for producing low interfacial tensions between hydrocarbons and water. J Colloid Interface Sci 59:31–38

26. Chai JL, Zhao JR, Gao YH, Yang XD, Wu CJ (2007) Studies on the phase behavior of the microemulsions formed by sodium dodecyl sulfonate, sodium dodecyl sulfate and sodium dodecyl benzene sulfonate with a novel fishlike phase diagram. Colloids Surf A 302:31–35

27. Chiang MY, Shah DO (1980) The effect of alcohol on surfactant mass transfer across the oil/brine interface and related phenomena. Presented at the SPE 5th International Symposium on oilfield and Geothermal Chemistry, Stanford, CA

28. Chilingar GV, Yen TF (1983) Some notes on wettability and relative permeability of carbonate rocks: II. Energy Resour 7:67–75

29. Comelles F, Pascual A (1997) Microemulsions with butyl lactate as cosurfactant. J Dispers Sci Technol 18:161–175

30. Dosher TM, Wise FA (1976) Enhanced oil recovery potential: an estimate. J Petrol Technol 28:575–585

31. El-Batanoney M, Abdel-Moghny T, Ramzi M (1999) The effect of mixed surfactants on enhancing oil recovery. J Surf Deterg 2:201–205

32. Elraies KA, Tan IM, Awang M, Fathaddin MT (2010) A new approach to low-cost, high performance chemical flooding system. Presented at the SPE Production and Operation Conference and Exhibition, Tunis, Tunisia

33. Engelskirchen S, Elsner N, Sottmann T, Strey R (2007) Triacylglycerol microemulsions stabilized by alkyl ethoxylate surfactants—a basic study. Phase behavior, interfacial tension and microstructure. J Colloid Interface Sci 312:114–121

34. Flaaten AK, Nguyen QP, Zhang J, Mohammadi H, Pope GA (2010) Alkaline/surfactant/polymer chemical flooding without the need for soft water. Presented at the SPE Annual Technical Conference and Exhibition, Denver, Colorado, USA

35. Gao B, Sharma MM (2013) A family of alkyl sulfate Gemini

surfactants. 2. Water–oil interfacial tension reduction. J Colloid Interface Sci 407:375–381

36. Glover FDS, Santanna VC, Barros Neto EI (1979) Surfactant phase behavior and retention in porous media. Soc Pet Eng J 19:183–193

37. Gurgel A, Moura MCPA, Dantas TNC, Barros Neto EL, Dantas Neto AA (2008) A review on chemical flooding methods applied in enhanced oil recovery. Braz J Pet Gas 2:83–95

38. Healy RN, Reed RL (1974) Physicochemical aspects of microemulsion flooding. Soc Pet Eng 14:491–501

39. Healy RN, Reed RL (1977) Immiscible microemulsion flooding. Presented at the SPE 4th symposium on improved oil recovery, Tulsa, OK

40. Healy RN, Reed RL, Carpenter CW (1975) A laboratory study of microemulsion flooding. Soc Pet Eng J 259:87–100

41. Healy RN, Read RL, Stenmark DG (1976) Multiphase microemulsion system. Soc Pet Eng J 261:147–160

42. Holm LW (1971) Use of soluble oils for oil recovery. J Pet Technol 23:1475–1483

43. Hussain A, Luckham PF, Tadros TF (1997) Phase behavior of ph dependent microemulsions at high temperatures and high salinity. Oil Gas Sci Technol Rev IFP 52:228–231 Iglauer S, Wu Y, Shuler P, Tang Y, Goddard WA III (2009) Alkyl polyglycoside surfactant–alcohol co-solvent formulations for improved oil recovery. Colloids Surf A 339:48–59

44. Iglauer S, Wu Y, Shuler P, Tang Y, Goddard WA III (2010) New surfactant classes for enhanced oil recovery and their tertiary oil recovery potential. J Pet Sci Eng 71:23–29

45. Jeirani Z, Mohamed Jan B, Si Ali B, Noor IM, See CH, Saphanuchart W (2013a) Formulation, optimization and application of triglyceride microemulsion in enhanced oil recovery. Ind Crop Prod 43:6–14

46. Jeirani Z, Mohamed Jan B, Si Ali B, Noor IM, See CH, Saphanuchart W (2013b) Formulation and phase behavior study of a nonionic triglyceride microemulsion to increase hydrocarbon production. Ind Crop Prod 43:15–24

47. John AC, Rakshit AK (1994) Phase behavior and properties of a microemulsion in the presence of NaCl. Langmuir 10:2084–2087

48. Kahlweit M, Strey R, Hasse D (1985) Effect of salt concentration on interfacial behavior in the surfactant system water ? noctadecane ? diethylene glycol monohexyl ether. J Phys Chem 89:163–167

49. Kahlweit M, Strey R, Busses G (1990) Microemulsions: a qualitative thermodynamic approach. J Phys Chem 94:3881–3894

50. Kassel DG (1989) Chemical flooding-status report. J Pet Sci Eng 2:81–101

51. Kayalia IH, Liub S, Miller CA (2010) Microemulsions containing mixtures of propoxylated sulfates with slightly branched hydrocarbon chains and cationic surfactants with short hydrophobes or PO chains. Colloids Surf A 354:246–251

52. Kim V, Hilfiker R, Eicke HF (1988) Excess adsorption of hydrocarbons on the oil/water interface in H2O/AOT/oil microemulsions in the presence of additives. J Colloid Interface Sci 121:579–584

53. Kumar R, Mohanty KK (2010) ASP flooding of viscous oils. Presented at the SPE Annual Technical Conference and Exhibition, Florence, Italy

54. Kwok W, Hayes RE, Nasi-El-Din HA (1995) Modeling dynamic adsorption of an anionic surfactant on Berea sandstone with radial flow. Chem Eng Sci 50:768–783

55. Lalanne-Cassou C, Carmone I, Fortney LN, Samii A, Schechter RS, Wade WH, Weerasoooriya U, Weerasoooriya V, Yiv S (1983) Minimizing cosolvent requirements for microemulsion formed with binary surfactant mixtures. J Dispers Sci Technol 8:137–156

56. Leung R, Hou MJ, Monohar C, Shah DO, Chun PW (1985) Microand microemulsions.In: Shah DO (ed). Am Chem Soc, Washington, DC

57. Levitt DB, Jackson A, Heinson C, Britton LN, Malik T, Dwarakanath V, Pope GA (2006) Identification and evaluation of highperformance eor surfactants. SPE Reserv Eval Eng 12:243–253

58. Li X, He G, Zheng W, Xiao G (2010) Study on conductivity

property and microstructure of TritonX-100/alkanol/n-heptane/ water microemulsion. Colloids Surf A 360:150–158

59. Lu J, Liyanage PJ, Solairaj S, Adkins S, Arachchilage GP, Kim DH, Britton C, Weerasooriya U, Pope GA (2014) New surfactant developments for chemical enhanced oil recovery. J Pet Sci Eng 120:94–101

60. Maerker JM, Gale WW (1992) Surfactant flood process design for Loudon. SPE Reserv Eng J 7:36–44

61. Maidment LJ, Chen V, Warr GG (1997) Effect of added cosurfactant on ternary microemulsion structure and dynamics. Colloids Surf A 129–130:311–319

62. Mandal A, Bera A, Ojha K, Kumar T (2012) Characterization of surfactant stabilized nanoemulsion and its use in enhanced oil recovery. Paper SPE 155406-MS, presented at International Oilfield Nanotechnology Conference and Exhibition, 12–14 June, Noordwijk, The Netherlands

63. Mclendon WJ, Koronaios P, Enick RM, Biesmans G, Salazar L, Miller A, Soong Y, McLendon T, Romanov V, Crandall D (2014) Assessment of CO_2-soluble non-ionic surfactants for mobility reduction using mobility measurements and CT imaging. J Pet Sci Eng 119:196–209

64. Meyers KO, Salter SJ (1980) The effect of oil brine ratio on surfactant adsorption from microemulsions. Presented at the SPE 55th annual fall technical conference and exhibition, Dallas, Texas Mo C, Zhong M, Zhong Q (2000) Investigation of structure and structural transition in microemulsion systems of sodium dodecyl sulfonate/n-heptane/n-butanol/water by cyclic voltammetric and electrical conductivity measurements. J Electroanal Chem 493:100–107

65. Myers D (2006) Surfactant science and technology, 3rd edn. Wiley, Hoboken

66. Nakamae M, Abe M, Ogino K (1990) The effects of alkyl chain lengths of sodium alkyl sulfates and n-alkanes on microemulsion formation. J Colloid Interface Sci 135:449–454

67. Nelson RC, Pope GA (1978) Phase relationship in chemical flooding. Soc Pet Eng J 18:325–338

68. Osborne DW, Middleton CA, Rogers RL (1988) Alcohol-free microemulsions. J Dispers Sci Technol 9:415–423

69. Osterloh WT, Jante MJ (1992) Surfactant-polymer flooding with anionic PO/EO surfactant microemulsions containing polyethylene glycol additives. Paper SPE 24151, presented at the SPE 8th Symposium on enhanced oil recovery, Tulsa, OK Pattarino F, Marengo E, Trotta M, Gasco MR (2000) Combined use of lecithin and decvl polyglucoside in microemulsions: domain of existence and cosurfactant effect. J Dispers Sci Technol 21:345–363

70. Pottmann FH (1974) Secondary and tertiary oil recovery process. Interstate Oil Compact Commission, Oklahoma City

71. Purwono S, Murachman B (2001) Development of non petroleum base chemicals for improving oil recovery in Indonesia. Paper SPE 68768, presented at the SPE Asia Pacific Oil and Gas Conference and Exhibition, Jakarta, Indonesia

72. Qiao W, Li J, Zhu Y, Cai H (2012) Interfacial tension behavior of double long-chain 1, 3, 5-triazine surfactants for enhanced oil recovery. Fuel 96:220–225

73. Qutubuddin S, Millar CA, Fort T (1984) Phase behavior of pHdependent microemulsions. J Colloid Interface Sci 101:46–58 Richardson CJ, Aboofazeli A, Lawrence MJ, Barlow JD (1997)

74. Prediction of phase behavior in microemulsion systems using artificial neural networks. J Colloid Interface Sci 187:296–303

75. Rosen MJ (1989) Selection of surfactant pairs for optimization of interfacial properties. J Am Oil Chem Soc 66:1840–1947

76. Sagitani H, Friberg SE (1980) Microemulsion systems with a nonionic cosurfactant. J Dispers Sci Technol 1:151–164

77. Salehi M, Johnson SJ, Liang JT (2008) Mechanistic study of wettability alteration using surfactants with applications in naturally fractured reservoirs. Langmuir 24:14099–14107

78. Santanna VC, Curbelo FDS, Castro Dantas TN, Dantas Neto AA, Albuquerque HS, Garnica AIC (2009) Microemulsion flooding for enhanced oil recovery. J Pet Sci Eng 66:117–120

79. Sayyouh MH (1994) Alkalines can improve miscibility of a surfactant–oil–brine system and areal sweep efficiency of oil. Oil Gas 1:13–17

80. Sayyouh MH, Abdel-Waly AA, George J, Salama AO (1991) Design of a microemulsion slug for maximizing tertiary oil recovery efficiency. SPE 20804

81. Schechter RS, Wade WH (1976) Research on tertiary oil recovery: annual report. University of Texas at Austin, Texas

82. Schramm LL, Stasiuk EN, Turner D (2003) The influence of interfacial tension in the recovery of bitumen by water-based conditioning and flotation of Athabasca oil sands. Fuel Proc Technol 80(2):101–118

83. Schulman JH, Stoeckenius W, Prince LM (1959) Mechanism of formation and structure of microemulsions by electron microscopy. J Phys Chem 63:1677–1680

84. Schwuger MJ, Stickdorn K, Plummer, MA, Roszelle, WO (1975) Patent number US3901317

85. Scriven LE (1976) Equilibrium bicontinuous structure. Nature 263:123–125

86. Shah DO (1981) Fundamental aspects of surfactant–polymer flooding process. Keynote paper presented at the European Symposium on Enhanced Oil Recovery, Bournemouth, England

87. Shah DO (ed) (1985) Macro- and microemulsions. ACS Symposium Series 272; American Chemical Society, Washington, DC

88. Shah DO, Schechter RS (1977) Improved oil recovery by surfactant and polymer flooding. Academic, New York

89. Sharma MK, Shah DO (1985) Macro- and microemulsion. In: Shah DO (ed). Am Chem Soc, Washington, DC

90. Shindy AM, Darwich TD, Sayyaouh MH, Abdel-Aziz O (1997) Development of an expert system for EOR method selection. Paper SPE-37708-MS, presented at the SPE Middle East Oil Show and Conference, 15–18 March, Bahrain

91. Southwick JG, Svec Y, Chilek G, Shahin GT (2010) The effect of live crude on alkaline-surfactant-polymer formulations: implications for final formulation design. SPE-135357-MS, presented at the SPE Annual Technical Conference and Exhibition, 19–22 September, Florence, Italy

92. Stoeckenius W, Schulman JH, Prince LM (1960) The structure of

myelin figures and microemulsionsa s observed with the electron microscope. Kolloid-Z 169:170–178

93. Trabelsi S, Argillier J, Dalmazzone C, Hutin A, Bazin B, Langevin D (2011) Effect of added surfactants in an enhanced alkaline/heavy oil system. Energy Fuels 25:1681–1685

94. Verkruyse LA, Salter SJ (1985) Potential use of nonionic surfactants in micellar flooding. Paper SPE 13574, presented at the SPE International Symposium on Oilfield and Geothermal Chemistry, Phoenix, Arizona

95. Wadle A, Fö̈rster Th, von Rybinski W (1993) Influence of the microemulsion phase structure on the phase inversion temperature emulsification of polar oils. Colloids Surf A 76:51–57

96. Wan T, Yao J, Zishun S, Li W, Juan W (2011) Solution and drilling fluid properties of water soluble AM–AA–SSS copolymers by inverse microemulsion. J Pet Sci Eng 78:334–337

97. Wang D, Liu C, Wu W, Wang G (2010) Novel surfactants that attain ultra-low interfacial tension between oil and high salinity formation water without adding alkali, salts, cosurfactants, alcohol and solvents. Paper SPE 127452, presented at the SPE EOR Conference at Oil and Gas West Asia, Muscat, Oman

98. Wei J, Huang G, Yu H, An C (2011) Efficiency of single and mixed Gemini/conventional micelles on solubilization of phenanthrene. Chem Eng J 168:201–207

99. Willhite GP, Green DW, Okoye DM, Looney MD (1980) A study of oil displacement by microemulsion systems-mechanisms and phase behavior. Soc Pet Eng J 20:459–472

100. Wilson PM, Murphy CL, Foster WR (1976) The effects of sulfonate molecular weight and salt concentration on the interfacial tension of oil–brine–surfactant systems. Paper SPE 5812-MS, presented at the SPE Improved Oil Recovery Symposium, 22–24 March, Tulsa, OK

101. Winsor PA (1954) Solvent properties of amphiphile compounds. Butterworth's Scientific Publications, London

102. Wormuth KR, Kaler EW (1987) Amines as microemulsion cosurfactants. J Phys Chem 91:611–617

103. Yangming Z, Huanxin W, Zulin C, Qi C (2003) Compositional modification of crude oil during oil recovery. J Pet Sci Eng 38:1–11

104. Zhang DL, Lui S, Yan W, Puerto M, Hirasaki J, Miller CA (2006) Favorable attributes of alkali–surfactant–polymer flooding. Paper SPE 99744, presented at the SPE symposium on improved oil recovery, Tulsa, OK

105. Zhou M, Rhue RD (2000) Effect of interfacial alcohol concentrations on oil solubilization by sodium dodecyl sulfate micelles. J Colloid Interface Sci 228:18–23

106. Zhu Y, Xu G, Gong H, Wu D, Wang Y (2009) Production of ultra-low interfacial tension between crude oil and mixed brine solution of Triton X-100 and its oligomer Tyloxapol with cetyltrimethylammonium bromide induced by hydrolyzed polyacrylamide. Colloids Surf A 332:90–97

Citations

CHAPTER 1

Ribeiro, F., Pedrosa, A. and Souza, M. (2014) Conversion of Isopropylbenzene over AlSBA-15 Nanostructured Materials. Modern Research in Catalysis, 3, 94-98. doi: 10.4236/mrc.2014.33012.

CHAPTER 2

Sachin Kumar and R. K. Singh, Thermolysis of High-Density Polyethylene to Petroleum Products, Journal of Petroleum Engineering, vol. 2013, Article ID 987568, 7 pages, 2013. doi:10.1155/2013/987568.

CHAPTER 3

T. Tugsuu, S. Yoshikazu, B. Enkhsaruul and D. Monkhoobor, A Comparative Study on the Hydrocracking for Atmospheric Residue of Mongolian Tamsagbulag Crude Oil and Other Crude Oils, Advances in Chemical Engineering and Science, Vol. 2 No. 3, 2012, pp. 402-407. doi: 10.4236/aces.2012.23049.

CHAPTER 4

AlemÃ¡n-Nava GS, Sandate-Flores L, Meneses-JÃ¡come A, DÃaz-Chavez R, Dallemand JF, et al. (2014) Bioenergy Sources and Representative Case Studies in Mexico. J Pet Environ Biotechnol 5:190. doi: 10.4172/2157-7463.1000190.

CHAPTER 5

Abdeen Z, El-Sheshtawy HK, Moustafa YMM (2014) Enhancement of Crude Oil Biodegradation by Immobilizing of Different Bacterial Strains on Porous PVA Hydrogels or Combining of them with their Produced Biosurfactants. J Pet Environ Biotechnol 5:192. doi: 10.4172/2157-7463.1000192.

CHAPTER 6

M. F. Attallah, N. S. Awwad and H. F. Aly (2012). Environmental Radioactivity of TE-NORM Waste Produced from Petroleum Industry in Egypt: Review on Characterization and Treatment, Natural Gas - Extraction to End Use, Dr. Sreenath Gupta (Ed.), ISBN: 978-953-51-0820-7, InTech, DOI: 10.5772/39223.

CHAPTER 7

Magne bråtveit, Jorunn kirkeleit, Bjørg eli hollund and Bente e. Moen, Biological Monitoring of Benzene Exposure for Process Operators during Ordinary Activity in the Upstream Petroleum Industry, doi: 10.1093/annhyg/mem029

CHAPTER 8

Albina Mukhametshina and Elena Martynova, Electromagnetic Heating of Heavy Oil and Bitumen: A Review of Experimental Studies and Field Applications, Journal of Petroleum Engineering, vol. 2013, Article ID 476519, 7 pages, 2013. doi:10.1155/2013/476519

CHAPTER 9

J. C. Tang, R. G. Wang, X. W. Niu, M. Wang, H. R. Chu, and Q. X. Zhou, Characterisation of the Rhizoremediation of Petroleum-contaminated Soil: Effect of Different Influencing Factors, doi:10.5194/bg-7-3961-2010.

CHAPTER 10

Abdullahi Dyadya Mohammed, Abubakar Garba Isah, Musa Umaru, Shehu Ahmed, Yababa Nma Abdullahi, Comparative Study on Sulphur Reduction from Heavy Petroleum - Solvent extraction and microwave irradiation approach, ISSN 2076-2909.

CHAPTER 11

Iman Najafi; Mahmood Amani, Asphaltene Flocculation Inhibition with Ultrasonic Wave Radiation: A Detailed Experimental Study of the Governing Mechanisms, DOI: 10.3968/j.aped.1925543820110202.108, ISSN 1925-5438.

CHAPTER 12

Albina Mukhametshina and Elena Martynova, Electromagnetic Heating of Heavy Oil and Bitumen: A Review of Experimental Studies and Field

Applications, Journal of Petroleum Engineering, vol. 2013, Article ID 476519, 7 pages, 2013. doi:10.1155/2013/476519.

CHAPTER 13

Achinta Bera and Ajay Mandal, Microemulsions: A Novel Approach to Enhanced Oil Recovery: A Review, J Petrol Explor Prod Technol DOI 10.1007/s13202-014-0139-5.

Index